MEN-AT-ARMS SERI

EDITOR: MARTIN WINDROW

The Canadian Army at War

Text and colour plates by MIKE CHAPPELL

OSPREY PUBLISHING LONDON

Published in 1985 by
Osprey Publishing Ltd
Member company of the George Philip Group
12–14 Long Acre, London WC2E 9LP
© Copyright 1985 Osprey Publishing Ltd

British Library Cataloguing in Publication Data

Chappell, Mike
    The Canadian Army at War.—(Men-at-arms
    series; 164)
    1. Canada. *Canadian Army*—History
    I. Title   II. Series
    355.3′1′0971   UA600

    ISBN 0-85045-600-2

Filmset in Great Britain
Printed in Hong Kong

Dedication
This book is dedicated to the memory of my uncle, Sergeant Gerald Bastable 12th Manitoba Dragoons (18th Canadian Armoured Car Regiment). Killed in action, Normandy, August 1944.

Acknowledgements
The author wishes to thank the following, who gave generously of their time and material during the compilation of this book:
Mr Barry Agnew of the Military History Department, Glenbow Museum, Calgary, Alberta; Brig.Gen. J. L. Summers (Retired), CMM, MC, CD, of Saskatoon, Saskatchewan; Mr René Chartrand, Chief, Military Section, Historic Sites Service, Parks of Canada; Regimental Adjutant, the Princess Patricia's Canadian Light Infantry; Mr David Ross of the Lower Fort Garry National Historic Park; and the Canadian Defence Liaison Staff, London.
    The main works consulted were:
The Canadian Army 1939–1945, Col. C. P. Stacey, OBE, AM, PhD, (King's Printer, Ottawa, 1948); *Canadian Expeditionary Force 1914–1919*, Col. G. W. L. Nicholson, CD (Queen's Printer, Ottawa, 1962); *Canada's Soldiers*, George F. G. Stanley (Macmillan, Toronto, 1960).

Canada and her Army

Looking back from the dawn of the present century, Canada can be said to have had a turbulent history. Frenchman, Briton, Indian and American revolutionary fought each other in a series of wars which ended with the prize of Canada in British hands. Even then Canada had to contend with periodic internal unrest and threats from beyond its frontiers. British troops were withdrawn in 1871 following Confederation, leaving the defence of Canada entirely in the hands of her own people. By the 1890s this task had been entrusted to a Militia force (said by some to be nothing more than a political gendarmerie), and a woefully small band of regulars.

Her troubled past had not, therefore, left Canada with a tradition of militarism. Militia and regular forces were never recruited up to establishment; expenditure on the Army was always hotly contested; and public opinion was split over the subject of defence policy. Isolationism had many supporters, strongly opposing Canadian involvement in what they saw as imperial adventures; whereas their opponents believed in alliance with Great Britain as a way to stay under her protective mantle. Woven into all argument was the question of the expansionist policy of the United States, seen as a threat by some, but welcomed by others who preferred the protection of the 'Monroe doctrine'.

It was in this see-saw atmosphere that Canada, not by any means a military country, was drawn into the first war of the 20th century—an episode that set in motion the sequence of events that was to lead her into two World Wars as well as a number of minor ones.

Armies are usually representative of the societies from which they are raised, and it may be argued that there was still much of the frontier spirit abroad in the Canada of the early 20th century. Still developing, it drew men seeking challenge and

Canadian Militia uniform, late 19th century. Cpl. Robert Moore of the 10th Royal Grenadiers pictured in the uniform he wore during the Riel Rebellion, 1885. Note the Havelock cap, full dress red tunic and corduroy trousers. His buff belts are scrubbed free of pipeclay, and a Snider .577 in. rifle completes his equipment. (Glenbow Archives, Calgary)

adventure, men who were hardy and able to think for themselves. That these ideal soldierly qualities were prevalent in Canada is borne out by the speed with which Canadian armies were formed, and by the tenacity and professional skill which they displayed when committed to battle. In the military history of the British Empire their achievements are legendary, ranking with the best and quite disproportionate to their size. All the more credit is reflected when it is remembered that Canada's expeditionary forces have always been found from volunteers.

The story of Canada and her Army in the 20th century is set out in the chapters that follow. It is a record of great effort and great achievement on the part of a nation which might justifiably have stayed aloof from the quarrels of others; but which, particularly in the two major wars of this century, chose to send her soldiers to fight for their liberty.

Canadian irregular uniform, late 19th century. One of the more colourful outfits raised at the time of the Riel Rebellion were the Rocky Mountain Rangers. No uniform was issued to the Rangers, and the dress of the man portrayed reflects his civilian calling. He is armed with a Government 1876-pattern Winchester carbine, supplemented by a revolver thrust into the waistband of his chaps. (Glenbow Archives, Calgary)

1900-14:
The Growing Commitment

In the period of tension leading up to the outbreak of the South African War, Britain looked to her Empire for aid. Opinion in Canada was sharply divided over the dispatch of Canadian troops, but eventually Sir Wilfred Laurier, the Prime Minister, gave way to the pro-war lobby and agreed to send a force of 1,000 volunteers to South Africa. With the precedent established, resistance to the active involvement of Canadian troops abated, allowing Britain's later call for more Canadian units to pass unchallenged.

The first Canadian contingent to arrive in South Africa was an infantry unit, the 2nd (Special Service) Battalion of the Royal Canadian Regiment. Though raw and virtually untrained on arrival in Cape Town, they were nevertheless able to take their place in the British 19th Brigade by the middle of February 1900. By 18 February they were part of the force investing the Boer positions at Paardeberg Drift. Eight days later they attacked as part of an operation which led to the surrender of Gen. Cronje and his army; and 27 February 1900 saw the first victory for the first Canadian troops to serve overseas. Paardeberg, which cost the 2nd RCR 34 dead and nearly 100 wounded, must rank as an extraordinary achievement for a unit composed largely of men who had been following their civilian occupations only $4\frac{1}{2}$ months previously.

Like the Canadian units that were to follow them to South Africa, the 2nd RCR was recruited from volunteers who had served in the Militia, with a small element of officers of the Permanent Force. The Militia, in one form or another, had been part of the Canadian military scene for centuries, and was now similar in many ways to the British Volunteer organisation. Service with the Canadian Militia at the end of the 19th century was voluntary, and involved a maximum of 16 days embodied training per year. Intended for home defence, the Militia would continue to provide a useful source of manpower for the expeditionary forces of the 20th century.

The Permanent Force at this time consisted of the

Royal Canadian Dragoons, a battalion of the Royal Canadian Regiment, and the Royal Canadian Artillery with a strength of two field batteries and two garrison companies. Formed as training staffs for the Militia, the Permanent Force had only recently taken titles suggesting a more active rôle.

As the 2nd RCR took part in the ordeal of marching and manoeuvre that led to the capture of Bloemfontein and Pretoria, the call went out for more Canadian units. This time, however, mounted troops were demanded, in an attempt to beat the Boer commando at his own game. In response the Royal Canadian Dragoons (the retitled 1st Canadian Mounted Rifles); the 2nd, 3rd, 4th, 5th and 6th Canadian Mounted Rifles (of which only the first two saw active service); Strathcona's Horse; a field brigade of the Royal Canadian Artillery, and a field hospital were raised, equipped, and sent out to Africa. With the contingents went Maj.Gen. E.

Sgt. Arthur Richardson, Strathcona's Horse, pictured after winning the VC in South Africa in July 1900. Richardson rode back into a withering Boer crossfire to rescue a fallen comrade.

Officers of the Royal Canadian Regt. photographed in South Africa, 1900. (Canadian War Museum)

5

T. H. Hutton, commander of the Canadian Militia, who took up command of the 1st Mounted Infantry Division.

In the hard campaigning that followed the initial British reverses, the Canadian troops won a reputation for toughness and valour. Criticism was voiced at their equipment and administration, but never at their performance in the field. At the head

Canadian Militia uniform, early 20th century. Pte. Charles Light, 22nd Saskatchewan Light Horse, pictured prior to the Great War. His tunic is scarlet, with a blue cap and breeches. Piping, cap band and shoulder straps are white. The Oliver-pattern waistbelt and the Ross rifle are typically Canadian. (Doug Light, Calgary)

of the list of awards to Canadians were four Victoria Crosses, one being awarded to the future Lt.Gen. Sir Richard Turner, then a lieutenant in the Royal Canadian Dragoons. Turner won his award at Leliefontein on 7 November 1900, when a small group of Royal Canadian Dragoons and Royal Canadian gunners held off 200 Boers. Lt. Cockburn and Sgt. Holland also won the supreme award during this action.

The defeat of the Boers put an end to Canada's first overseas war, but not to the wrangling at home over imperial defence. Canadian troops in South Africa had not enjoyed the total support of their countrymen, and those opposed were determined to see that the South African experience would not be repeated. The British proposal at the 1902 Colonial Conference that Canada should contribute a force of 4,500 men to an Imperial Reserve was rejected, resulting in a more subtle approach at the conference of 1907. The British line was now to point out the desirability of standardisation in organisation, weapons, equipment and training. This was conceded, along with the concept of an Imperial General Staff with a 'local section' 'advising' the Canadian Government on military matters.

In 1909 the Imperial Defence Conference again stressed standardisation; but also agreed on a British mission to Canada to structure the General Staff, and on the training of Canadian officers in Britain. Another outcome of the conference was the preparation of mobilisation schemes including, significantly, one for a Canadian expeditionary force.

After the next, and secret, meeting of an Imperial Defence committee in 1911 (which included the soon-to-be legendary Col. Sam Hughes) Canada was effectively bound to the line of Imperial Defence which her leaders had so flatly refused nine years earlier. Thenceforward, wherever Great Britain chose to venture militarily, Canada had to follow.

While these negotiations were under way, the Militia of Canada was undergoing a thorough overhaul to make it as effective a force as possible. Disagreement with a series of British generals (including the abrasive Hutton of the pre-Boer War period) led to the abolition of the office of GOC Canadian Militia, and its replacement by a Militia Council. Under this body the organisation and training of the Militia was brought up to date, new weapons and equipment were brought into service (including the controversial Ross rifle), and the building of additional military installations proceeded. With the world moving inexorably towards a European conflict, Canada stood prepared for her part. Her military leaders, like those of all the future participants, believed that the war would be a brief affair—a fairly bloodless combat in which superior manoeuvre would decide the day, and from which the spoils awaiting the victors justified the risks.

1914-18: Attrition

On 2 August 1914 Germany demanded the right to march her armies through Belgium in order to attack France. Belgium refused and appealed to Great Britain for aid. Britain warned Germany not to violate Belgian neutrality, but on the morning of 4 August German troops crossed the Belgian frontier. Her ultimatum rejected, Britain found herself at war with Germany.

Behind Britain stood her Empire, and one by one the member nations rallied to the aid of the 'old country'. On 6 August the offer of troops made by the Canadian premier, Sir Robert Borden, was accepted, and the order went out for the immediate mobilisation of a Canadian expeditionary force.

Raising an Army

A plan for this eventuality had stood ready for many years. Awaiting its implementation were the Canadian Militia, Permanent Force and General Staff; but the Minister of Militia, Sam Hughes, had ideas of his own. Having decided that the mobilisation scheme was inadequate and time-wasting this extraordinary man scrapped it. Virtually single-handed, he then set about the organisation of the biggest army Canada had ever put into the field.

In a whirlwind of activity, Hughes began by ordering officers commanding Militia units to select volunteers and to form them into numbered units of an expeditionary force—units with little connection with the existing Militia structure. The following examples may serve to illustrate typical expeditionary force units:

The 16th Battalion drew its first company from the 50th Gordon Highlanders of Canada; its second from the 72nd Seaforth Highlanders of Canada; its third from the 79th Cameron Highlanders of Canada; and its fourth from the 91st Canadian Highlanders. Each group wore its own tartan and badges for many months until uniform dress and insignia were agreed upon and procured. The '3rd Provisional Battalion' was formed from 400 volunteers from the Royal Grenadiers, nearly 1,000 from the Queen's Own Rifles of Canada, and 40 from the Governor General's Bodyguard.

Next, at great expense and effort, Hughes ordered a large mobilisation centre to be built as rapidly as possible at Valcartier, Quebec. Here the expeditionary force assembled; drew stores, equipment and transport; and began training for the great task ahead.

Amid the gathering of new and rootless units was one which had been formed as a private venture. Andrew Hamilton Gault, a Militia captain who had served with the Canadian Mounted Rifles in South Africa, had sought out Sam Hughes in the weeks prior to the outbreak of war and had offered to raise and equip a unit at his own expense. Hughes found the proposition attractive and, once mobilisation was under way, gave Gault permission to proceed with his plan. In conjunction with Lt.Col. Farquhar, the military secretary of the Governor-General, Gault raised an infantry battalion of over 1,000 men who had previously served in the British

Men of the 1st Bn., Canadian Expeditionary Force on Salisbury Plain, December 1914. Note the 'Canadian'-pattern Service Dress (quite different from the British pattern), and the General Service maple leaf badge that preceded battalion and regimental badges. (IWM Q53262)

armed forces. The battalion was named the Princess Patricia's Canadian Light Infantry (after the daughter of the Governor-General), and was recruited in nine days. Lt.Col. Farquhar commanded, and Andrew Hamilton Gault became its senior major. The PPCLI were to be the first Canadian infantry to see action, crossing from England to France on 21 December 1914, and entering the line with the British 27th Division on 4 January 1915.

On 3 October 1914 the Canadian Expeditionary Force sailed for England in one of the largest convoys of ships and escorts ever assembled up to that time. Over 30,000 strong, their numbers included the British 2nd Lincolnshire Regt., who had been relieved as the garrison of Bermuda by the Royal Canadian Regiment. The feelings of Canada's only regular infantry at this ignominious turn of events can be imagined, but they were not to

Pte. Howard Exton and comrades of the 31st Bn., CEF, parade at Calgary in 1915. Note the Canadian-pattern Service Dress, Ross rifles and Oliver-pattern equipment rigged for marching order. (Glenbow Archives, Calgary)

remain in the 'second team' for long. The steady stream of men and munitions that flowed from Canada over the war years soon carried them in the wake of the first contingent, to England and then on to the Western Front.

Training in England

Following their arrival in Plymouth the Canadian Expeditionary Force went to Salisbury Plain to prepare themselves for France. This was one of the wettest winters on record; their three months of intensive training would have been arduous enough in the tented and temporary camps they were allocated, but the extreme weather made it a misery. It was a testing time for the Canadians, but they came through it with great fortitude.

Not only the weather was a taste of things to come. At least one Canadian unit on Salisbury Plain became infected with lice; survivors of the experience were to remember with relish how their commanding officer, outraged at the lapse of discipline, paraded his unit to threaten that he would not take 'one lousy man to France'!

On the completion of training the Canadian Expeditionary Force separated to form an infantry division and a cavalry brigade, while hospitals and sundry other units began to leave to join British formations. (The PPCLI had already departed; and it would be some time before it rejoined the Canadian fold. Another battalion to depart were the Newfoundlanders, who had travelled to England with the Canadians and now joined the British 29th (Regular) Division to win an enviable reputation at Gallipoli, and to be virtually wiped out on the first day of the Somme. They remained with the British Army throughout the war, finishing up with the 9th (Scottish) Division on occupation duties in Germany.) Depôts were formed at Tidworth and Devizes; a Royal review was held on 4 February 1915; and three days later the Canadian Division left for Avonmouth to board the ships which were taking them to France.

Three further Canadian divisions were to follow them. The 2nd, after staging and training around Shorncliffe, Kent, went to France in September 1915. The 3rd was assembled in France between December 1915 and March 1916; and the 4th crossed to France in August 1916, after concentration in the Aldershot Command.

Ypres 1915: The First Battles

After a period in a quiet part of the line, by April 1915 the Canadian Division had moved to the Ypres Salient. Here, on the 22nd of the month, the Germans launched an offensive aimed at the capture of Ypres, and the Canadians were subjected to a new and terrifying instrument of war—poison gas. Following a brisk artillery bombardment, the Germans released 160 tons of chlorine gas into a favourable wind which bore the deadly vapour towards the lines of the Canadian Division and its flanking formations. Surprise was total. Unprotected, the Allies suffered the full horror of gas attack.

The ordeal was too much for the French 87th (Territorial) and 45th (Algerian) Divisions: they broke, leaving a gap of four miles to the left of the Canadians. Into this the Germans advanced, and then dug in; the Canadians responded fiercely, containing the threat and buying time to enable the Allied line to be re-established.

On 24 April the Germans repeated their plan. Once more a vicious artillery bombardment was followed by the release of gas. The full brunt was

Group of Canadian troops, June 1916. At least two wear the badge of the 15th Bn. (48th Highlanders of Canada). The remainder of the insignia and uniform is a motley selection typical of the soldier at the front.

borne by the Canadians north-west of St Julien; driven back under a hail of fire, sick and choking, they resisted stubbornly, keeping the enemy in check until they were relieved. Canada's citizen soldiers had acquitted themselves well in their first trial, but at enormous cost. Over 6,000 casualties had been sustained in the four days of fighting.

Severe casualties were also suffered by the 'Princess Pat's', fighting further to the south with the British 27th Division. In the battle for Bellewaarde Ridge on 8 May the PPCLI, now commanded by Maj. Hamilton Gault, lost nearly 400 men in the fight to maintain their positions. They came out of the line with only 150 effectives.

Brought up to strength once more, the Canadian Division saw action in the Artois offensive at Festubert in May and at Givenchy in June. Whatever success the Canadians may have been able to claim in this ill-conceived operation was once more offset by the casualties sustained—nearly 3,000 dead, wounded and missing.

It was at this time that the division's Ross rifles were officially withdrawn and replaced by British Lee-Enfields. The Ross had proved unreliable in the muddy conditions of the Western Front because of the tendency of its 'straight-pull' bolt mechanism to jam. By officially replacing it authority was merely sanctioning the actions of the front-line Canadians, who scrounged the rifles of British dead and wounded. (A few Ross rifles were kept for sniping, where their superior accuracy was an advantage.)

The Canadian Corps

With the arrival at the Front of the 2nd Canadian Division a Canadian Corps was formed. Command was assumed by Lt.Gen. E. Alderson, the British officer who had commanded the original Canadian Division from its formation. The 2nd Division was commanded by Maj.Gen. R. E. W. Turner, the South African War vc; and Brig.Gen. Arthur

Currie was promoted major-general to command the original division, now the 1st.

The formation of the Corps enabled a number of Canadian units which had been serving elsewhere to be brought back under Canadian command. An independent infantry brigade included the PPCLI back from the 27th British Division; and the Royal Canadian Regiment, returned from its exile in Bermuda. There was also an infantry brigade consisting entirely of Canadian Mounted Rifles units. The strength of the Corps in September 1915 totalled nearly 38,000 officers and men.

Over the winter of 1915/16 the newly-formed Corps took responsibility for a sector of the Front between Ploegsteert Wood and St Eloi. Here the time passed relatively quietly, if uncomfortably, until the prelude to the great Somme offensive.

The Somme 1916: The Big Push

On 1 July 1916, after almost two years of building and training, the British high command launched their new armies into a major offensive. Hopes were high for this first really big push by the British, but disaster befell the venture from its first day, when a

Capt. Buchanan and men of the 13th Bn. (Royal Highlanders of Canada) in a front-line trench, July 1916; a photo full of contrast and curiosity. Note the US Army pistol belt on the officer on the left, Capt. Buchanan's automatic pistol (just visible in the holster), and the total lack of uniformity.

staggering 60,000 casualties were sustained for minimal gains. For the next five months the sordid trade-off in lives and material continued. At the close of the offensive 600,000 British casualties had been suffered, to exact 435,000 casualties from the Germans. The maximum penetration on the 30-mile Front was seven miles.

Initially, the Canadian Corps was not directly involved at the Somme. Holding a sector of the line south of Ypres, they were engaged in the battle for the St Eloi craters in April, when the 2nd Division suffered nearly 1,400 casualties; and in the battle of Mount Sorrel in June, when a German offensive succeeded, resulting in the loss of much ground which had to be retaken. Here the Corps sustained 8,000 casualties, including the commander of the newly-activated 3rd Division, Maj.Gen. M. S. Mercer, who was killed in the opening moments of his division's first battle.

A belated casualty of the fight for the St Eloi craters was the Corps commander, Gen. Alderson. Having seen the Canadian Expeditionary Force through a most difficult period, he was considered to be 'incapable of holding the Canadian Divisions together'. Moved to a post in England, he was succeeded in May by Lt.Gen. Sir Julian Byng, under whom the fortunes of the Canadian Corps were to prosper.

At about this time the Ross rifle, still the official weapon of the 2nd, 3rd and 4th Divisions, was finally replaced by the Lee-Enfield. Fiercely championed by the Minister of Militia, the Ross had remained the standard arm of the Canadians for nearly two years despite its very obvious limitations (not the least of which was that it would only accept Canadian-made ammunition without jamming). The Colt machine guns of the Canadians were also replaced by British Vickers guns, by now becoming available in sufficient quantities.

In August 1916 the Canadian Corps left Flanders and marched south to the Somme battlefield. On the sleeves of their uniforms they now wore the 'battle patches' that would identify them for the remainder of the war (and also in the war of 1939–45). The 1st Division wore a red patch; the 2nd, dark blue; the 3rd, black—changed to French grey later; and, on their arrival, the 4th wore green. Brigades, battalions, headquarters, etc. were identified by supplementary patches worn above, or

The Western Front, 1915–18; the principal theatre of Canadian operations in the Great War. The shaded area indicates the terrain fought over for most of the war.

sometimes on, the divisional patch.

Taking over from the Australians at Pozieres in early September, the Canadians began preparations for their own offensive. On 15 September the Canadian Corps attacked as part of a two-army operation. Their objective was the village of Courcelette, a key point on the left flank of the assault. The battle differed from those that had preceded it in two important respects. First, tanks were employed to support the infantry; and secondly, the recently-evolved artillery tactic of the 'creeping barrage' was used. (Troops pressing hard on the heels of a creeping barrage were exposed to less risk than when a barrage ceased moving altogether, or switched a great distance on their approach to the objective.)

In the din of massed artillery, and accompanied by the clattering tanks, the 2nd and 3rd Canadian Divisions went 'over the bags' at 6.20 am. Under the terrible weight of the barrage the German forward defences were smashed and occupied within 15 minutes. Pushing on, the 2nd Division reached their first objectives in just over an hour. The six tanks accompanying the Canadians failed early in the operation, only one reaching its objective; but their presence struck terror into the enemy.

Now it was the turn of the 5th and 7th Brigades to pass through the first wave and assault Courcelette. This fortified town was entered by the 22nd (Canadien Français), 25th and 26th Battalions, who fought for three days to clear it and to repel the many German counter-attacks which soon followed.

On the left the 42nd (Royal Highlanders of Canada) Battalion and the PPCLI distinguished themselves in a difficult fight to secure the flank; but by now the Germans had recovered, and the fighting grew in intensity as they sought to regain the ground lost to them. To add to the difficulties the weather now broke; but the Canadians consolidated their gains and stabilised the line.

The fighting of 15 to 22 September had cost the Corps over 7,000 casualties, and had won them the praise of the Commander-in-Chief who recorded that, 'The result of the fighting of 15 September and following days was a gain more considerable than any which had attended our arms in the course of a single operation since the commencement of the offensive.'

A Canadian nurse photographed at Etaples in 1917. Canada did not follow the example of the British in forming women's auxiliary units in the Great War, but Canadian nurses served in many theatres of war including the Western Front and the Middle East.

Now began a bloody struggle for the possession of Thiepval Ridge and the Ancre Heights. From 26 September until 8 October the Canadian Corps participated in a series of assaults that met with little or no success and served only to swell the casualty lists. The objective that had eluded capture— Regina Trench—finally fell to the 4th Canadian Division (operating independently from the Canadian Corps) on 11 November at their third attempt. At the end of that month the 4th Division handed over its sector of the Somme and marched to join its parent Corps on the Lens-Arras Front.

The Somme fighting established the reputation of the Canadians as a corps d'élite respected by friend and foe. Resolute, experienced, and well-led, they had come through an experience that had shattered other formations, and they were to move on to even greater achievements. The cost had been high, however: over 24,000 Canadians had been killed, wounded or posted as missing in the three months of struggle in Picardy.

Vimy Ridge, 1917: Canada's Triumph

Allied strategic planning for the spring of 1917

Sentries of the 22nd Bn. (French Canadians) in a trench, July 1916. The trench is clean and well-maintained and the men are alert, properly dressed and at their posts. Note the Ross rifle and bayonet. A good example of a well-disciplined unit on active service.

called for a major effort from the French Army. Gen. Nivelle's offensive was to be supported by a concerted thrust by the British, and much faith was placed in the ability of the victor of Fort Douaumont to repeat his Verdun success on a grander scale.

The Germans, however, had plans of their own. In February they began falling back between Soissons and Arras to a heavily-fortified position which both shortened their line and conformed to a new concept of defence in depth—the Hindenburg Line. The Somme bloodbath of the previous year had caused the Germans to question their tactics. Already holding the strategic advantage, they now sought a new tactical advantage in concrete pillboxes, massive wiring, and reserves who were kept well back from the enemy fire.

By the time the Allies had fought their way up to the Hindenburg Line the French offensive was already behind schedule. When it was eventually set in motion on 16 April the Germans had not only observed the preparations, but had obtained copies of the outline plan and certain operation orders. One month's fighting by the French gained but four miles, and at a cost of over 100,000 casualties. In May, with the French Army in a virtual state of mutiny, Nivelle was sacked. His successor, Gen.

Petain, was faced with the unenviable task of restoring his army's morale while trying to contain the enemy. To the great good fortune of the Allies, the German High Command never fully appreciated the situation the French were in; and pressure elsewhere kept them occupied for the remainder of 1917.

Further north, the British had launched an offensive at Arras on 9 April, in which the task allocated to the Canadian Corps was the capture of Vimy Ridge. Since 1914 all previous attempts to take the ridge had failed. The Germans considered it a vital anchor to their defence of the area, and had fortified it to such a degree that it was considered impregnable. From their supposedly unassailable position they dominated the ground below. No movement in daylight went unobserved, and their gunners ruled the battlefield. The German position was constructed on the 'old' system of rigid trench lines, however, and by the spring of 1917 little had been done to implement their 'new' in-depth order.

The staffs of the Canadian Corps realised that the

Canadian cavalry watering their horses, Comblain l'Abbé, 1917. This unit is the Canadian Light Horse, by this time the Corps Cavalry Regiment. Each squadron wore the badges of the unit from which it came. The majority visible here are from the 19th Alberta Dragoons, the remainder from the Royal North-West Mounted Police. Oddly enough, the Albertas wear the stetson hats!

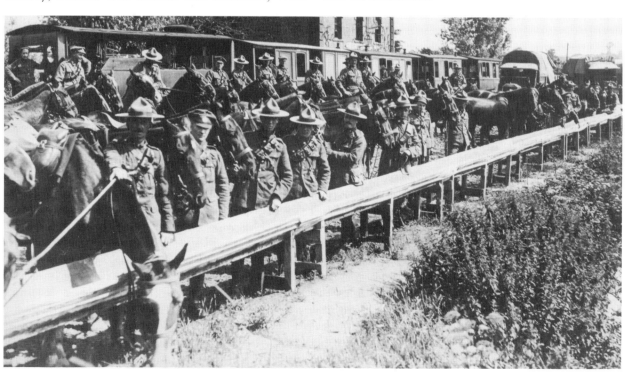

Lt.Gen. Sir Arthur Currie, GCMB, KCB, commander of the Canadian Corps from 1917 until the end of the Great War.

A captain of the 2nd Canadian Motor Machine Gun Bde., April 1918. Note the arrow patch on his shoulders (an arrow was the symbol for a machine gun on marked maps). This officer wears the badges of the Yukon Battery, Canadian Machine Gun Corps and the ribbon of the Military Cross.

battle for Vimy Ridge would have to be won by artillery before the infantry assault was launched. They assembled the Corps artillery, substantially reinforced by British heavy artillery, in an arc facing the German position. Some 245 heavy and over 600 field pieces were positioned, with over 200 additional guns of 1st British Corps and 1st Army on call. Tens of thousands of tons of ammunition were dumped ready for the two-week bombardment that would precede the assault.

This would be no simple slogging match, however. The available firepower was to be used in ways that were both ingenious and novel. Counter-battery fire, a relatively new concept at the time, was to be used to neutralise the enemy artillery. To this end aerial observation, aerial photographs and sound-ranging were used to discover the enemy gun positions on the reverse slopes—over 80 per cent were thus located and plotted. An elaborate communication system was set up to enable fire control to be as flexible as possible: by day observed

fire would concentrate on enemy strongpoints, while harassing fire—including that from machine guns—would hit the rear areas day and night. A new fuse was used with high explosive shell to cut wire more effectively than before; and on the day of the assault a rolling barrage would precede the infantry while standing barrages continued to bombard the enemy beyond. All these elaborate artillery plans were co-ordinated by the Canadian Corps' chief gunner, Brig.Gen. E. W. B. Morrison.

While the gunners prepared, the sappers and the infantry toiled. To move the huge quantities of ammunition and stores needed, roads and small-gauge railways had to be built and maintained. Water pipelines were laid and 21 miles of telephone cable was dug in (in addition to the 60 miles of cable laid on the ground). With the area under constant surveillance by the enemy, the bulk of the work was done at night.

Not detected by the enemy were the 'tunnellers'. These troops laboured to dig, or to extend, miles of tunnels and caverns, through which reliefs and reinforcements could pass in safety and unobserved. (Many of these tunnels still exist, and visitors to the Vimy Memorial Park are today led through them by young Canadian volunteer guides, obviously proud of the achievements of their countrymen so long ago.)

Since most of these preparations were unavoidably conspicuous, the Canadians made no attempt to conceal their intentions from the Germans. They let them know they were coming; what they concealed was *when*. In the rear areas mock-ups of the battle area were constructed on which detailed rehearsals were conducted. Nothing was left to chance, and every man knew the part he had to play in the battle to come.

On 20 March the artillery bombardment began. At first only part of the available batteries were employed, but on 2 April the full might of the Canadian and British guns fell upon the German lines and rear areas. Nothing like it had been seen before. Under the deluge of high explosive the German lines crumbled into a waterlogged moonscape, as the Canadian infantry probed and patrolled nightly. Finally, knowing that the enemy would expect an intensification of fire before the assault, the bombardment was slackened as the clock ticked towards zero hour.

At 5.30 am on Easter Monday, 9 April 1917, the attack on Vimy Ridge started with the concerted roar of the guns supporting the infantry assault—21 Canadian battalions representing all four divisions of the Corps, and fighting together for the first time. In driving snow and sleet, and keeping well up to their barrage, the heavily-laden infantry picked their way to the German Front line with minimal opposition. From here resistance began to stiffen, but the assault was pressed home on schedule. By 6.25 am most of the battalions had reached their first objectives.

On the Canadian left flank progress was slow, but on the right the 1st and 2nd Divisions pressed on in great style, clearing the ridge to reach their final objectives before the end of the day. In places the Germans resisted stubbornly, but the artillery preparation and supporting barrages had done their work. Compared to the Somme battles the infantry's task was an easy one. Only the 4th Division on the extreme left had a difficult time; but they too secured their objectives by 12 April.

It had been a resounding victory. Driven from the ridge they considered impregnable, the Germans were forced to fall back, as much as four miles in places, from the high ground now in the

A major of the 22nd Bn., June 1918. Officially known as the '22ᵉ Canadien Français' at this time, the 22nd were the forerunners of the famous 'Vandoos' (i.e. 'Vingt-Deux', '22'). Note the cap and collar insignia; the cravat worn in place of the regulation collar and tie; the divisional insignia on his arms (red circle over a blue rectangle); and the gold insignia (CII) worn on the divisional patch by all officers of the 2nd Division. The ribbon is the Military Cross.

possession of the Canadians, who had captured 54 guns, 104 mortars, 124 machine guns and over 4,000 prisoners. Well might Arthur Currie, whose 1st Division had performed so well, write of 'the grandest day the Corps has ever had'. But Vimy Ridge was not taken without cost: over 10,000 Canadian casualties were sustained in its capture, of whom 3,598 died.

Four VCs were earned on Vimy Ridge, and a knighthood for Maj.Gen. Currie (soon to succeed Gen. Byng as Corps Commander). But perhaps the greatest accolade was bestowed by the French: they sent a General Staff study group to the scene of the Canadian victory to review the battle. One of the greatest armies in the world was not too proud to learn from an army of citizen-soldiers.

Ypres 1917: Passchendaele

With the French still in disarray it became necessary for the British to maintain the pressure begun at Arras and Vimy Ridge. The Canadian

15

Corps continued to operate near the scene of its recent triumph, while an offensive was planned for Flanders in mid-summer. After a successful operation at Messines in June, the British attacked from Ypres in July. The usual intensive artillery bombardment had smashed what was left of the drainage system on the low-lying ground; and the heavy rain that began as the operation got under way had turned the battlefield into a quagmire. Into this impossible situation waded the British.

Meanwhile the Canadians kept up offensive operations in their sector, culminating in the successful assault on Hill 70 in August, and the subsequent fighting for the city of Lens. In October the Canadian Corps was ordered north to the Ypres salient. They relieved the exhausted Australians on the same sector of the Front which the original Canadian contingent had held in April 1915. However, the survivors of that time could find little that was familiar in the pulverised and sodden terrain now confronting them. Faced with the task of taking the village of Passchendaele, they had first to overcome the appalling ground conditions in order to bring up the guns and munitions necessary to make the assault feasible.

Men of the Princess Pat's Light Infantry rest after battle, 31 August 1918, at Arras. Note the 'battle patches' of a green semicircle over the French-grey rectangle of the 3rd Div., the PPCLI titles, and cap badge worn on the helmet by the man in the right foreground.

On 26 October the Canadians attacked and, despite heavy casualties, succeeded in gaining a lodgement on Bellevue spur, one of the key pieces of higher ground. Four days later they attacked again, once more gaining ground with heavy losses as they fought their way nearer to the rubble of Passchendaele. Finally, on 6 November, they entered the ruins of the village, the honour of being the first troops to enter falling to the 27th (City of Winnipeg) Battalion.

It was the end of the campaign that would one day be called 'Third Ypres' or simply the battle of Passchendaele. Its grand strategic aim was never realised, but it prevented the Germans from exploiting the sad state of the French Army. British losses from July to November totalled almost a quarter of a million, including the 16,000 Canadian troops killed, wounded and missing in the effort to break through at Passchendaele.

1918: Spearheading the Victory

After being relieved in the salient the Canadian Corps returned to Vimy. Here they prepared for the German offensive expected in the spring of 1918. The collapse of the Russians had released large numbers of German formations from the Eastern Front, and these were being marshalled for a decisive battle.

When the blow came it fell upon the British 5th

Army, far to the south of the Canadian sector. On 21 March the Germans struck the perilously extended British, and drove them back across the ground won so dearly at the Somme. For the next four months the Germans struck repeatedly at the Allied line and made spectacular gains; but their main aim, that of driving a wedge between the British and the French, was never realised. By the summer the Germans were exhausted. The time was now ripe for the Allies to strike back.

After an elaborate deception plan the Canadian Corps, in concert with Australian and French formations, struck suddenly at Amiens on 8 August. The hoped-for surprise was achieved and, spearheaded by tanks, the Corps advanced 12 miles in three days as German resistance crumbled. The German Gen. Ludendorff was to comment, '8 August was the black day of the German Army in the history of this war.' Having got the enemy on the run, Lt.Gen. Currie wanted to press on; but the British General Staff had other ideas. After years of trying to rupture the enemy front, they became nervous when presented with an unexpected breakthrough. Advance on a broad front was decided upon, and the impetuous Canadians were sent to Arras with orders to assault the Hindenburg Line there.

In hard fighting between 26 August and 9 October the Canadian Corps once more cut their way through a series of supposedly impregnable German positions. Always in the vanguard, they breached the Hindenburg Line and the Canal du Nord fortifications, and captured the infamous Bourlon Wood and the city of Cambrai. The Germans then began a general withdrawal, and fighting reverted to the open warfare of the early months of the war. With this achieved, the Allies harried their enemy into the phase of war so long striven for—pursuit.

The Germans were beaten. Their morale, both at home and at the Front, was low; their leaders were considering armistice in order to prevent the ignominy of total defeat; and units of their armed forces were on the point of mutiny. But there were men among them who were prepared to fight and die to check the Allied advance; and they enabled the German Army in France and Flanders to fall back in some semblance of order. Against this opposition the Canadians fought their way through

Sgt. Walter Rayfield, VC, 7th Bn. CEF. Rayfield won the supreme award at Arras in September 1918. His cap badge bears the battalion's secondary title '1st British Columbia', and the 'battle patch' is all red.

Valenciennes to reach Mons by 11 November. It was here, with the Armistice concluded, that the bugles sounded the ceasefire.

In the last 100 days of the war the Canadian Corps had consistently advanced over distances which made those gained in the great and bloody offensives of the Somme and Passchendaele seem trivial. So, too, did the numbers of prisoners taken, and the guns and machine guns captured. But this had not been achieved without sacrifice. Right up until the closing moments of the war the Canadian Corps paid the price in lives and suffering of being a vanguard. The arithmetic speaks for itself. Canada was then a nation of just eight million people. Of these over 600,000 served in the armed forces, of whom over 60,000 gave their lives. Nearly 140,000 were wounded or gassed.

Canada's major contribution to the victory was undoubtedly made by the Canadian Corps under the leadership of Lt.Gen. Sir Arthur Currie; but it had not been the only contribution. Fighting with the British 3rd Cavalry Division was the Canadian Cavalry Brigade, which included the Royal

A Canadian brigadier-general examines a German Bergmann MP 18, August 1918. The brigadier's brassard has the 'patch' of the 4th Div. (in which all officers wore a gold maple leaf on the green patch) sewn to it. Another officer wears the brassard of the Canadian Corps.

Canadian Dragoons, Lord Strathcona's Horse, the Fort Garry Horse and the Royal Canadian Horse Artillery. Canadians also served in specialist units concerned with the operation of railways, cutting and clearing timber, tunnelling and medical duties. Combatant units saw service in Russia, fighting German and Communist forces; and a brigade of Canadians served for a time in eastern Russia.

Now, with the war over, Canada's citizen soldiers returned home. After a token march to the Rhine, the Canadian Corps broke up as shipping became available to take its units back home, where they received the victor's welcome they so richly deserved.

Major Units and Appointments of the Canadian Corps, November 1918

General Officer Commanding—Lt.Gen. Sir Arthur Currie, GCMG, KCB.

1st Canadian Division (GOC Maj.Gen. Sir A. C. MacDonell, KCB, CMG, DSO)

1st Bde. Canadian Field Arty.; 2nd Bde. CFA; 1st Bde. Canadian Engineers

1st Inf. Bde.: 1st, 2nd, 3rd & 4th Bns. CEF

2nd Inf. Bde.: 5th, 7th, 8th & 10th Bns. CEF

3rd Inf. Bde.: 13th, 14th, 15th & 16th Bns. CEF

1st Bn. Canadian Machine Gun Corps

1st Divisional Train, RCASC

1st, 2nd and 3rd Field Ambulances

2nd Canadian Division (Maj.Gen. Sir H. E. Burstall, KCB, CMG)

5th Bde. Canadian Field Arty.; 6th Bde. CFA; 2nd Bde. Canadian Engineers

4th Inf. Bde.: 18th, 19th, 20th & 21st Bns. CEF

5th Inf. Bde.: 22nd, 24th, 25th & 26th Bns. CEF

6th Inf. Bde.: 27th, 28th, 29th & 30th Bns. CEF

2nd Bn. Canadian Machine Gun Corps

2nd Divisional Train, RCASC

4th, 5th and 6th Field Ambulances

3rd Canadian Division (Maj.Gen. F. O. W. Loomis, CB, CMG, DSO)

9th Bde. Canadian Field Arty.; 10th Bde. CFA; 3rd Bde. Canadian Engineers

7th Inf. Bde.: Royal Canadian Regt., PPCLI,

42nd & 49th Bns. CEF
8th Inf. Bde.: 1st, 2nd, 4th & 5th Bns. Canadian Mounted Rifles
9th Inf. Bde.: 43rd, 52nd, 58th & 116th Bns. CEF
3rd Bn. Canadian Machine Gun Corps
3rd Divisional Train, RCASC
8th, 9th and 10th Field Ambulances

4th Canadian Division (Maj.Gen. Sir D. Watson, KCB, CMG)
3rd Bde. Canadian Field Arty.; 4th Bde. CFA; 4th Bde. Canadian Engineers
10th Inf. Bde.: 44th, 46th, 47th & 50th Bns. CEF
11th Inf. Bde.: 54th, 75th, 87th & 102nd Bns. CEF
12th Inf. Bde.: 38th, 72nd, 78th & 85th Bns. CEF
4th Bn. Canadian Machine Gun Corps
4th Divisional Train, RCASC
11th, 12th and 13th Field Ambulances

Company sergeant major of the 16th Bn. (Canadian Scottish), marching order, Great War period. (IWM Q30226)

Commanding officers of the battalions comprising the 3rd Canadian Inf. Bde., 1918. From left to right: 14th Bn. (Royal Montreal Regiment); 15th Bn. (48th Highlanders of Canada); 16th Bn. (Canadian Scottish); and 13th Bn. (Royal Highlanders of Canada, the Canadian Black Watch). Lt.Col. C. W. Peck of the Canadian Scottish was a holder of the Victoria Cross—see Plate C3.

Corps Troops
Canadian Field Arty.; Canadian Garrison Arty.; Canadian Engineers; 1st and 2nd Motor Machine Gun Bdes. Canadian Machine Gun Corps; RCASC; RCAMC; 1st to 13th Bns. Canadian Railway Corps; Canadian Labour Corps; Canadian Forestry Corps.

1939-45: *The Global War*

To the generation who had fought in the Great War of 1914–18, the prospect of yet another war was unthinkable. It had been 'the war to end wars'. Great efforts were made in the 1920s and 1930s to prevent conflict, including the setting up of the ill-fated League of Nations, while armies and spending on defence were cut to the bone. In Canada it seemed as if national defence had become a very low priority. Many of the past reasons for the maintenance of a defence force were no longer valid by the 1920s, and public opposition to spending on armaments was strong.

In this climate the Militia and the Permanent Force were restructured—given new titles, and precious little else. It was a time of military stagnation in Canada and elsewhere, and the depression that struck the world in 1929 served only to make nations more concerned with their domestic problems—too concerned to pay much attention to the territorial smash-and-grab of the Japanese and Italians. The activities of Hitler in the late 1930s at last awakened the free world to the inevitable. Nations began to re-arm, hoping as they did so that a war might somehow be averted.

In Canada the Militia was reorganised in 1936 to enable it to fight a modern war, while the first steps were taken to procure the arms and equipment it would need. Canada's Army would fight to defend Canada's neutrality, although plans were made for a small expeditionary force. This, however, would not leave Canada's shores without the approval of Parliament.

Canadians in Britain

When Britain finally declared war on Hitler's Germany on 3 September 1939 Canada was not bound, as in 1914, to follow. A partial Canadian mobilisation was ordered, and a state of emergency

declared. On 7 September the Canadian Parliament began a two-day debate which ended with general approval of the Government's motion for war. On 10 September Canada declared war on Germany.

Immediately, plans were put into effect for the organisation of the Canadian Active Service Force, as the expeditionary force was now to be called. The assembly of the CASF, once again a volunteer force, was attended by none of the chaotic improvisation of the 1914 mobilisation. All went according to plan as Government and General Staff deliberated as to where, and in what strength, the CASF might be used. Eventually it was decided to send a division to Great Britain, and on 17 December 1939 the 1st Canadian Infantry Division landed at Greenock, Scotland, from whence they travelled south to Britain's largest military centre, Aldershot.

This sprawling brick-and-slate garrison, half-town, half-barracks, would become—with its satellite camps—a Canadian Army centre for the rest of the war. Situated 35 miles south-west of London, it was an unlovely place. The drabness of its mid-Victorian buildings was, however, alleviated by its tree-lined avenues, acres of sportsfields and surrounding countryside; and it

Toronto Scottish drilling at Aldershot, late 1939. In this 'phoney war' period much of the weaponry and equipment issued to the Canadians was of Great War vintage. (IWM H590)

was a big improvement on the conditions that the 1914 CEF had endured on Salisbury Plain.

Here the Canadians served out the period of the 'phoney war', until the German blitzkrieg, launched on 10 May 1940 and leading to the surrender of France on 22 June, changed the status of the Canadians completely. From a token force, marking time in a backwater, they became—after the débâcle of Dunkirk—a front-line division, one of the few with its equipment and morale intact. With Britain now in a state of siege, and with invasion expected any day, the 1st Canadian Infantry Division stood by for the desperate battle to come. (Some Canadians were sent to Brittany in the closing days of the Battle for France. The 1st Canadian Infantry Brigade landed in Brest on 13 June as part of a forlorn attempt to retain a foothold on the Continent; but they were evacuated after the French capitulation.)

For the next three years the 1st Canadian Infantry Division and the formations which followed them to England stood on guard. These were as follows: 2nd Canadian Infantry Division in 1940; 3rd Canadian Infantry Division, 1st Canadian Army Tank Brigade and 5th Canadian Armoured Division in 1941; 4th Canadian Ar-

Canadians at Spitzbergen. In late August 1941 a small force of Canadians removed the population of Spitzbergen and destroyed the installations and facilities there. These men are from the covering force, the Edmonton Regiment. (IWM H13618)

moured Division in 1942; and 2nd Canadian Army Tank Brigade in 1943.

Canadians in the Pacific

On the other side of the world were two battalions destined to be the first Canadians to see action. Late in 1941 the Royal Rifles of Canada and the Winnipeg Grenadiers were sent to augment the tiny British garrison of Hong Kong. They were exchanging one garrison duty for another, as one battalion had formerly been in Newfoundland and the other in Jamaica. Their commander was Brig. J. K. Lawson.

Scarcely had the Canadians settled into their new surroundings than the Japanese struck at Pearl Harbor on 7 December, and at Hong Kong the following day. The British general commanding the 14,000-strong defence force had deployed the Canadians on Hong Kong island itself, but in the early stages of the battle 'D' Company of the Winnipeg Grenadiers became the first Canadian soldiers to fight when they were sent to the mainland to reinforce the British brigade there. Shortly afterwards the British withdrew to the island.

After an intensive bombardment and a series of air strikes, the Japanese began landing on Hong Kong island on the night of 18 December. The fighting went on until Christmas Day, by which time the situation had become hopeless for the defenders. Pocket by pocket they were taken out,

Another photograph of the 'phoney war' period. Still at Aldershot, the Toronto Scottish practise anti-aircraft drills with the newly-issued Bren Light Machine Gun. Note the Canadian battledress: of a darker, greener shade than the British item, its quality and cut were superior.

Canadians in Sicily, July 1943. 'D' Co. of the Carleton and York Regt. (3rd Bde./1st Div.) march inland from the beaches. Their respirators and additional gear will soon be jettisoned. (IWM NA4491)

fighting desperately and exacting a heavy toll of the Japanese. Brigadier Lawson died, pistol in hand, when his headquarters was overrun; and nearly 800 of the 1,975 Canadians on Hong Kong became casualties in the fighting. What was left of the garrison surrendered to spend the remainder of the war in Japanese prisons. The initial death toll of 290 was later swelled to 554 by the privations they had yet to suffer.

In August 1943 Canadian troops once more set out to do battle with the Japanese; this time in the Aleutian Islands, which the Japanese had occupied in 1942. The Allied force set up to retake the Aleutians was a joint American/Canadian venture, totalling 34,000 men of whom 4,800 were Canadians of the 13th Infantry Brigade (including the re-formed Winnipeg Grenadiers and the Rocky Mountain Rangers). Canadians also formed part of the dual-nationality 1st Special Service Force, destined to spearhead the operation. The troops landed on 15 August only to discover that the Japanese had evacuated the island two weeks earlier. What would certainly have been a bloody battle ended as an anti-climax.

A Canadian Army Pacific Force was organised after the defeat of the Germans in 1945. Its main component was designated the 6th Canadian Division, but its employment was rendered unnecessary by the surrender of Japan after the dropping of the atomic bombs on Hiroshima and Nagasaki.

August 1942: Dieppe

In Britain, by the summer of 1942 the threat of invasion had receded. The air war preceding invasion had been won by the British in 1940, and Hitler had turned his attention to Russia. The forces in Britain had increased in numbers, efficiency and equipment, and reinforcements were beginning to arrive from the United States. The Canadian Army in England had grown first to Corps dimensions, and then to an Army. It was fully trained and equipped, and anxious to get into the fight. Apart from minor operations the Canadians had not been tested. Their chance was to come in the summer of 1942.

At this time Allied opinion was divided regarding strategic aims. America and Russia were in favour of an invasion across the English Channel, the so-called 'Second Front', while British interest was centred on North Africa and the Mediterranean.

(An ardent champion of the Second Front invasion was the Canadian, Lord Beaverbrook, who used his newspaper and his cabinet position to broadcast his views.)

British raids on the mainland of Europe had been occurring since the Dunkirk evacuation. From being mere pinpricks they had grown more and more ambitious, culminating in the desperate affair at St Nazaire in March 1942. Combined Operations Headquarters now decided to mount a raid in divisional strength: the target selected was the port of Dieppe, and the formation chosen to hit it was the 2nd Canadian Infantry Division.

The plan for the Dieppe raid was British, as were the officers in overall command of the operation. The Canadian Army was presented with this plan, which seemed feasible in its original concept, and was invited to carry it out. After two-and-a-half years of waiting for action the Canadians jumped at the chance, whatever their misgivings. When bad weather resulted in the cancellation of the operation in July they were bitterly disappointed. From their embarkation ports the troops of the 2nd Division dispersed, only to find that the operation had been revived and scheduled for 19 August. By now, the original plan had been significantly altered. The aerial bombardment and the parachute assaults, scheduled to take place before the seaborne attack, had been eliminated. Very heavy reliance was now to be placed on surprise.

Surprise, however, was lost after the left wing of the ships carrying the attackers to Dieppe ran into a German flotilla at 3.47 am on 19 August. With the vital element of the plan gone, the operation degenerated into a major disaster. Only on the right flank, where Lord Lovat's No. 4 Commando carried out a copy-book attack, was there any success. The South Saskatchewans and the Camerons of Canada also managed to get ashore on this flank and, despite heavy resistance which prevented them from reaching their objectives, gave a good account of themselves. The remainder of the force

A Canadian Sherman tank thunders through a Sicilian town, July 1943. (IWM 5591)

Canadians in Italy: the Seaforths of Canada hold a memorial service at Ortona early in 1944. Note the regimental titles and divisional patch. (IWM NA10979)

met with the full fire-power of the alerted Germans, which turned the beaches into a slaughter-house. Despite the bravery shown that day, few Canadians had the opportunity to hit back at the enemy who rained fire on them, most being condemned to remain in whatever shelter they could find. By mid-morning the full impact of the disaster had been appreciated by the commands afloat, and withdrawal was ordered.

In the aftermath of the raid the phrase 'Reconnaissance in Force' was coined. Dieppe, it was later claimed, pointed up valuable lessons which were put into practice in the great Normandy invasion. Whether Canadian casualties at Dieppe were sustained in the interests of reconnaissance, or to silence those clamouring for a Second Front, is even now open to bitter argument. Beyond argument, however, are the casualty figures: 5,000 of the 6,000 troops at Dieppe were Canadian; of these a little over 2,000 returned, of whom 1,000 had never got ashore. Of the 3,000 Canadians left behind at Dieppe, over 900 were dead and the remainder prisoners. There were, therefore, few to report the findings of the 'reconnaissance'. But all—survivors, dead and prisoners—could bear witness to the futility of a Second Front at that time.

Canadians in the Mediterranean

By the spring of 1943, with the tide of war now flowing in favour of the Allies, Canadian public opinion was becoming impatient with what it saw as the inactivity of its Army in England. Impatient too was the 1st Canadian Army itself. It wanted to fight, and the schemes to provide battle experience for a small number of Canadians in Tunisia were regarded merely as a sop. It was decided, therefore, to 'blood' a substantial Canadian contingent in the forthcoming Allied invasion of Sicily. The formations chosen were the 1st Canadian Infantry Division and the 1st Canadian Army Tank Brigade; Maj.Gen. G. G. Simonds was to command.

At dawn on 10 July 1943 the Canadians landed near Pachino, at the southernmost tip of Sicily. On the left flank of the British 8th Army, they were the link with the United States 7th Army. Pressing inland, the Canadians at first encountered only light opposition, but the going was difficult and opposition began to stiffen as the enemy used the mountainous terrain to good advantage. A series of progressively bloodier battles led to the five-day fight for the town of Agira, which fell to the Canadians on 28 July, the honour of taking the town itself going to the PPCLI. In early August,

Sgt. Chanette of the 12th Canadian Armd. Regt. (Three Rivers Regt.) takes a time check over the radio of his Sherman; Italy, 1944. Note the red-on-black patch of the 1st Canadian Army Tank Brigade. (IWM NA11450)

1: Sgt., 90th Winnipeg Rifles; Canada, 1885
2: Pvt., 2nd Bn. Royal Canadian Regt.; S.Africa, 1900
3: Sgt., Strathcona's Horse; S.Africa, 1900

A

1: Sgt., Princess Patricia's Canadian Lt.Inf.; England, 1914
2: Machine gunner, 1st Canadian Contingent; England, 1915
3: Cavalry trooper, 1st Canadian Contingent; England, 1915

B

1: Sgt.,73rd Bn. CEF, 1915
2: RSM,15th Bn. CEF, 1918
3: Lt.Col.,16th Bn. CEF, 1918

C

1: Trooper, 19th Alberta Dragoons; England, 1916
2: Driver, Canadian Field Artillery, 1918
3: Sgt.,4th Bn. CEF, 1918

D

1: Lt.Col.,Royal Canadian Regt.,1926
2: Canadian infantryman, Hong Kong garrison, 1941
3: Canadian infantryman, Aleutian Islands, 1943

E

1: Cpl., Fusiliers Mont-Royal; Dieppe, 1942
2: Maj., Royal 22e Regt.; Italy, 1944
3: Sgt., CWAC; England, 1945

F

1: Sgt., Royal Winnipeg Rifles; 'D-Day', 1944
2: Sgt., 12th Manitoba Dragoons; Normandy, July 1944
3: Brig., 9th Inf.Bde., 3rd Canadian Div.; NW Europe, 1945

G

1: Sgt.,Royal Canadian Regt.;Korea, 1953
2: Guardsman, Canadian Guards; UN Forces, Cyprus,1964
3: Cpl.,PPCLI;'Canadian Forces' uniform, 1971

H

after more difficult fighting in the mountains west of Mount Etna, the Canadians were ordered into reserve and took no further part in the fighting for the island.

Having been given the first real opportunity to show their mettle, the Canadians had done well—the campaign had led to the liberation of Sicily in just 38 days. They had harried a determined enemy over 150 miles of mountain road and beaten him whenever he chose to fight. The battlefield experience so eagerly sought had been gained, but at the cost of over 2,000 casualties. The Canadians of the 1st Infantry Division and the 1st Army Tank Brigade now prepared themselves for the invasion of Italy.

As the 8th Army (with the Canadians in the vanguard) crossed the Straits of Messina on 3 September 1943, the Italian government was in the act of surrendering. The Germans, however, had no intention of giving up Italy. Further Allied assaults at Salerno and Taranto forced them to conduct a rapid fighting withdrawal from southern Italy; and they fell back to a defensive line constructed across the peninsula with its hinge on Cassino, a mountain feature barring the way to Rome.

Landing at Reggio the Canadians pushed

Men of the PPCLI after the battle that broke the Hitler Line; Italy, 1944. Note the titles, divisional patches, Canadian-pattern khaki drill bush shirts and denim trousers. Medical orderlies, these men stand before their stretcher-jeeps. (IWM NA15470)

inland, brushing aside resistance to take Potenza, a key centre east of Salerno, in 17 days. With the subsequent breakout from the Salerno bridgehead the Allies closed on the German stronghold, their progress slowing as the enemy fought back with growing tenacity.

As winter set in the 8th Army struck at the line of the Sangro River while the US 5th Army approached Cassino. On the Adriatic coast the Canadians fought their way to Ortona, where a savage battle for the town raged over the Christmas of 1943. The seven-day house-to-house contest was finally decided when a Canadian flanking attack threatened to cut off the German main position. On the night of 27/28 December the Germans pulled out of Ortona, conceding victory to the Canadians. The severity of the weather now put an end to offensive operations and made the holding of the line a miserable and wretched undertaking.

With the recent arrival of the 5th Canadian Armoured Division there were now over 75,000 Canadian troops in Italy, organised as the 1st Canadian Corps and commanded by Lt.Gen. H. D. G. Crerar.

Over on the west coast the Allies had been battering away at the Cassino position, and by the spring of 1944 had managed to beat down the first of its defences, the Gustav Line. Behind it lay another just as formidable, the Hitler Line; and against the centre of this, on 23 May, the 1st Canadian Corps launched an attack. Troops of the 1st Canadian Infantry Division breached the line after heavy fighting, and the tanks of the 5th Canadian Armoured Division rolled through the gaps to fight through and take up the pursuit. By 31 May the road to Rome was open. Unhappily, the Canadians were withdrawn into reserve at this juncture, and the honour of entering the Eternal City went to the forces of the United States.

Once more the Germans fell back upon a prepared position, this time the Gothic Line in northern Italy; and in late August the 1st Canadian Corps began its attack in the eastern part of this formidable position, at Pesaro. The task was the capture of Rimini. In a month's hard fighting through the hills above Rimini the Canadians battled with the Germans and the weather to seize the high ground dominating the town, from which they could see the plains of northern Italy. But by

Piper of the Seaforth Highlanders of Canada; Italy, 1944. (IWM NA11569)

now the rains had turned the rivers into torrents and the low-lying land into marshes. After weeks of bitter fighting the Canadians and their Allied formations reached the line of the Senio River and here, in January 1945, the front once again 'stabilised' for the winter.

This was the end in Italy for the 1st Canadian Corps; in February it was withdrawn from the theatre and sent to Holland to rejoin the 1st Canadian Army. The division of this Army had led to controversy and the resignation of the first Army Commander; now, for the closing months of the war, all the Canadian formations were to fight as one.

Sicily and Italy had cost Canada dear. Over 25,000 casualties had been sustained in the 20 months of fighting, of whom nearly 6,000 had died.

Canadians in North-West Europe

'D-Day', 6 June 1944, saw the long-awaited Allied invasion of Nazi-occupied Europe. Under the cover of total air supremacy over 6,000 vessels, from mighty battleships to tiny assault landing craft, delivered five infantry divisions, three armoured brigades, Commandos and Rangers to the beach-head in Normandy. On the flanks, three airborne divisions went into action by parachute and glider to secure vital points. With a lodgement established, further Allied formations poured ashore.

The ensuing battle decided the fate of France. Drawn on to the anvil of Caen by the British and Canadians, German forces in Normandy were encircled by an American break-out, and trapped and defeated at Falaise in August. With the Allied invasion of southern France the Germans withdrew towards their national borders, while the Allies pursued them on a 'broad front'.

Spearheading the Canadian contribution to this vast undertaking were the 3rd Canadian Division, the 2nd Canadian Armoured Brigade and (as part of the British 6th Airborne Division) the 1st Canadian Parachute Battalion. Having fought their way ashore on 'Juno' beach, the Canadians were pitched into some of the most savage fighting of the war as the Germans resisted stubbornly on the outskirts of Caen. Progress was slow, but a beachhead was established and into this came the build-up of men and munitions necessary for the break-out. By the time Caen fell in early July the 2nd Canadian Corps had arrived in Normandy, and on the 23rd of the month the headquarters of the 1st Canadian Army became operational. This formation included the 2nd and 3rd Canadian Infantry Divisions, the 4th Canadian Armoured Division and the 1st Polish Armoured Division among its major components. (It also had under command at various times British, American, Belgian and Dutch troops.)

Now began the drive for Falaise, launched on 25 July, the day the United States forces broke through at St Lô to start the manoeuvre which threatened the German forces in lower Normandy with encirclement. Given the task of closing the door to the 'Falaise pocket' the Canadians fought a series of battles that led to the entry of the town on 16 August, from whence the 4th Canadian and the 1st Polish Armoured Divisions pushed on in an attempt to link up with the Americans.

Realising their true situation, the Germans now made frantic efforts to escape the trap that was about to close. While the Canadians and Poles fought to contain them, they were mercilessly bombarded by Allied air and ground forces as the battle for the Falaise pocket drew to a close. At least eight German divisions had been destroyed, and about twice that number had been severely mauled. German armoured formations had used their fire-power and mobility to break through encirclement,

but the carnage in the pocket was horrific: 12,000 prisoners surrendered to the Canadian Army alone.

Regrouping, the Allies pursued the retreating Germans. The Canadians, being given the left flank, had the task of clearing the Channel coast. On 23 August a rapid advance was begun in the

Canadians in Normandy. A carrier of the Royal Hamilton Lt. Inf. escorts German prisoners of war to the rear during the fighting of August 1944.

face of stiff opposition, but by 26 August the Seine was reached, and Rouen was taken on the 30th after a three-day battle.

With the crossing of the Seine the Battle of Normandy was concluded. It had been a resounding victory for the Allies. In manpower alone the Germans had sustained losses estimated at 400,000, whilst their losses in materiel were staggering. By early September, with most of France liberated, it seemed as if the war must surely be over. But the decision to pursue the war against Germany to unconditional surrender had long been taken; unable or unwilling to unseat their Nazi masters, the German people chose to fight on.

Driving on, the 1st Canadian Army fought its way along the Channel coast, liberating town after town and overrunning the launching sites from which the Germans were bombarding south-east England with 'Vengeance Weapons'. (Ironically, the 2nd Canadian Infantry Division arrived at Dieppe prepared for a bloody battle, only to discover that it had been evacuated by the enemy.) By late September the Canadians were at the Scheldt estuary, the sea approach to the great port of Antwerp.

While the Germans retained control of the Scheldt estuary, Antwerp—now of vital importance to the Allies—could not be used as a supply port. The Canadians were now ordered to clear the Germans from their positions commanding the estuary. Operations began on 1 October and lasted until 8 November. These were weeks of hard fighting under extremely difficult conditions as the Canadians, with the British troops under their command, strove to dislodge the enemy from the region, which was in places too water-logged for movement yet not navigable for small boats. On 28 November, however, with the last mines cleared from the port's approaches, the first Allied supply convoy steamed into Antwerp; and for the three months following the Canadians held the line along the River Maas from Nijmegen.

With the failure of the Allied thrust into northern Germany in the autumn, the ill-fated Arnhem operation, and the unexpected German offensive through the Ardennes in December 1944, it was not until February 1945 that the Allies were once again ready to take the offensive.

By this time the 1st Canadian Army had grown to enormous proportions due to numerous Allied formations being brought under command. For the closing months of the war Gen. Crerar directed an army that at times numbered 13 divisions, and mustered over a third of a million men. For the battle of the Rhineland, which lasted through February and into March, the 1st Canadian Army comprised more British formations than Canadian; but all shared in the savage fighting through the Reichswald Forest, the breaking of the Siegfried Line, and the clearing of the Hochwald Forest. By 10 March the enemy had been driven east of the Rhine.

On 23 March the Allies crossed the Rhine, with troops of the 3rd Canadian Infantry Division and the 1st Canadian Parachute Battalion participating in this great operation. Henceforward the main Canadian effort was to be in Holland, however, where the 1st Canadian Army—now with its 1st Corps back from Italy, and fighting for the first time as a truly Canadian Army—was given the task of completing the liberation of that country, and of driving from Holland into northern Germany.

The Canadians spent the closing weeks of the war pursuing a beaten enemy and fighting the fanatics amongst them who chose a last-ditch stand sooner than surrender. On 28 April the Germans in western Holland agreed to a truce to enable

Gen. Montgomery with Gen. Crerar (right) and Maj. E. J. Brady of the Lincoln and Welland Regt. (4th Canadian Armd. Div.) in NW Europe, 1944. Maj. Brady had just received the ribbon of his immediately awarded Distinguished Service Order. (IWM B14875)

urgently needed supplies to be delivered to the starving civil population. On 4 May came the ceasefire that preceded a series of surrender ceremonies as the German forces finally laid down their arms. It was all over. The unconditional surrender so bitterly fought for had at last been forced upon the German nation. The cost was dear, especially so among the ranks of the Canadian volunteers. Since the landings in Normandy nearly 48,000 Canadian casualties had been sustained, of which 11,546 were fatal.

With the war in Europe over, the Canadian Army in that theatre rapidly dispersed. An occupation contingent at divisional strength was maintained in Germany for a year, and large numbers of men volunteered for service in the Pacific; but the majority of Canada's soldiers returned home, to resume their civilian lives. Canada's effort in the Second World War had been considerably greater, both in manpower and production, than in the war of 1914–18. More than one million men and women had enlisted in the Canadian armed forces, and over 45,000 had laid down their lives in the cause of other peoples' freedom. This, from a nation of 11 million people, was an admirable effort.

1st Canadian Parachute Bn., NW Europe, 1944/45. Operating with the British 6th Airborne Div., the Canadian Parachute Bn. participated in the Normandy landings and the Rhine crossing. In the latter operation Cpl. F. G. Topham (pictured above), a medical orderly with the Canadian paras, won the VC for rescuing wounded under fire. Note Canadian Paras' cap badge worn on the maroon beret, and Canadian parachutist's brevet worn on the left breast.

Major Units and Appointments of the 1st Canadian Army, 1945

General Officer Commanding—Gen. H. D. G. Crerar, CH, CB, DSO

General Officer Commanding 1st Canadian Corps—Lt.Gen. C. Foulkes, CB, CBE, DSO

General Officer Commanding 2nd Canadian Corps—Lt.Gen. G. G. Simonds, CB, CBE, DSO

Army Troops: 25th Armd. Delivery Regt. (The Elgin Regt.); The Royal Canadian Arty.; The Royal Montreal Regt.

1st Corps Troops: 1st Armd. Car Regt. (The Royal Canadian Dragoons); The Royal Canadian Arty.

2nd Corps Troops: 18th Armd. Car Regt. (12th Manitoba Dragoons); The Royal Canadian Arty.

1st Canadian Infantry Division (Maj.Gen. H. W. Foster, CBE, DSO)

4th Reconnaissance Regt. (4th Princess Louise Dragoon Guards); The Royal Canadian Arty.; The Saskatoon Light Inf. (MG)

1st Infantry Brigade
The Royal Canadian Regt.
The Hastings and Prince Edward Regt.
The 48th Highlanders of Canada

2nd Infantry Brigade
The Princess Patricia's Canadian Light Inf.
The Seaforth Highlanders of Canada
The Loyal Edmonton Regt.

3rd Infantry Brigade
Le Royal 22ᵉ Regt.
The Carleton and York Regt.
The West Nova Scotia Regt.

2nd Canadian Infantry Division (Maj.Gen. A. B. Matthews, CBE, DSO, ED)

8th Reconnaissance Regt. (14th Canadian Hussars); The Royal Canadian Arty.; The Toronto Scottish Regt. (MG)

NW Europe: HM King George VI decorates Brig. J. M. Rockingham, CBE, DSO, commander of the 9th Bde., 3rd Canadian Infantry Division. Standing on the right is Lt.Gen. Simonds. Brig. Rockingham went on to command the 25th Canadian Bde. in Korea. (IWM B10798)

4th Infantry Brigade
The Royal Regt. of Canada
The Royal Hamilton Light Inf.
The Essex Scottish Regt.
5th Infantry Brigade
The Black Watch of Canada
Le Regiment de Maisonneuve
The Calgary Highlanders
6th Infantry Brigade
Les Fusiliers Mont-Royal
The Queen's Own Cameron Highlanders of Canada
The South Saskatchewan Regt.
3rd Canadian Infantry Division (Maj.Gen. D. C. Spry, DSO, & Maj.Gen. R. H. Keefler, CBE, DSO, ED)
7th Reconnaissance Regt. (17th Duke of York's Royal Canadian Hussars); The Royal Canadian Arty.; The Cameron Highlanders of Ottawa (MG)
7th Infantry Brigade
The Royal Winnipeg Rifles
The Regina Rifle Regt.
1st Canadian Scottish
8th Infantry Brigade
The Queen's Own Rifles of Canada
Le Regiment de la Chaudiere
The North Shore (New Brunswick) Regt.
9th Infantry Brigade
The Highland Light Inf. of Canada

The Stormont, Dundas and Glengarry Highlanders
The North Nova Scotia Highlanders
4th Canadian Armoured Division (Maj.Gen. C. Vokes, CBE, DSO)
29th Armd. Reconnaissance Regt. (The South Alberta Regt.); The Royal Canadian Arty.
4th Armoured Brigade
21st Armd. Regt. (The Governor General's Foot Guards)
22nd Armd. Regt. (The Canadian Grenadier Guards)
28th Armd. Regt. (The British Columbia Regt.)
The Lake Superior Regt. (Motor)
10th Infantry Brigade
The Lincoln and Welland Regt.
The Algonquin Regt.
The Argyll and Sutherland Highlanders of Canada
5th Canadian Armoured Division (Maj.Gen. B. M. Hoffmeister, CB, CBE, DSO, ED)
3rd Armd. Reconnaissance Regt. (The Governor General's Horse Guards); The Royal Canadian Arty.
5th Armoured Brigade
2nd Armd. Regt. (Lord Strathcona's Horse)
5th Armd. Regt. (8th Princess Louise's (New Brunswick) Hussars)
9th Armd. Regt. (The British Columbia Dragoons)
The Westminster Regt. (Motor)
11th Infantry Brigade
The Perth Regt.
The Cape Breton Highlanders
The Irish Regt. of Canada
1st Canadian Armoured Brigade
11th Armd. Regt. (The Ontario Regt.)
12th Armd. Regt. (Three Rivers Regt.)
14th Armd. Regt. (The Calgary Regt.)
2nd Canadian Armoured Brigade
6th Armd. Regt. (1st Hussars)
10th Armd. Regt. (The Fort Garry Horse)
27th Armd. Regt. (The Sherbrook Fusiliers)
The Corps of Royal Canadian Engineers; The Royal Canadian Corps of Signals; The Royal Canadian Army Service Corps; The Royal Canadian Army Medical Corps; The Royal Canadian Ordnance Corps; The Royal Canadian Electrical and Mechanical Engineers; The Royal Canadian Provost Corps; Postal, Pay and Intelligence units.

1950-53: United Nations Involvement

On 25 June 1950 the Communist North Korean state attacked its South Korean neighbour, starting a war of three years' duration in a land that was primitive, hostile of climate, and remote to the countries of the United Nations who rallied to the cause of the South.

Korea had been divided along the line of latitude 38°N at the end of the Second World War, with Russian occupation in the north and that of the United States in the south. Thus hostile ideology was nurtured in the divided peninsula, flaring into war within five years.

While powerful North Korean forces were driving across the 38th Parallel the UN Security Council called for assistance from member nations to restore the situation. The first, and strongest, contribution was made by the United States, who had troops deployed by July. Shortly afterwards Britain and the Commonwealth responded, followed later by token contingents from 16 other countries.

Within weeks the UN and South Korean forces had contained the North Korean invasion (albeit from a toehold at Pusan) and, after a brilliant amphibious operation at Inchon, had reconquered South Korea to pursue the defeated aggressors north.

On 7 August Mr St Laurent, the Canadian Prime Minister, announced the decision to form a Canadian Army Special Force, to be trained and equipped to carry out Canada's obligations to the UN appeal.

The Special Force was raised as part of the Canadian Army Active Force (Canada's Regular Army) and was recruited from volunteers who were veterans, members of the Reserve Forces or civilians. The response to the call for men was good, and by the end of August the initial rush to enlist was over. It had not been without its problems, however, and many 'undesirables' had been drawn into the fold, including one man aged 72 and one with an artificial leg! Months were to pass before the problems created by hasty recruiting were overcome. The Commander of the Force, which was to be titled the 25th Canadian Infantry Brigade Group, was Brig. J. M. Rockingham.

The original intention was to train the Force up to unit level and then send it to Okinawa for collective training. As American ships were to be used, Seattle, USA, was the port of embarkation and the adjacent Fort Lewis the staging camp. However, by now operations in Korea were going well for the UN Forces who, by the middle of October, had driven the North Korean Army back across the border and were heading north for the Yalu River. In the light of this new development it was now recommended that the Canadian contribution be reduced to one battalion and used for occupation duties only. The remaining units of the CASF were to train in Fort Lewis.

When the 2nd Battalion, the Princess Patricia's Canadian Light Infantry, commanded by Lt.Col. Stone, left Seattle on 25 November, the war in Korea appeared to be nearly over. However, by the time the troopship arrived in Yokohama on 14 December the situation had completely changed. Chinese Communist Forces had intervened in North Korea, surprising the UN Command; and the occupation rôle which the 2nd PPCLI had been sent to fulfil no longer existed. Instead they were to be sent into action as soon as possible to help stop the advance from the north. This was contained by the end of January, and after intensive training the 'Patricias' moved to the 9th (US) Corps area, where they came under command of the 27th British Commonwealth Infantry Brigade on 17 February 1951.

The Canadians' first battles were fought during

Staghound armoured cars of the 12th Manitoba Dragoons (18th Armoured Car Regt.)—Reconnaissance Regt. of II Canadian Corps—in the Hochwald Forest, 1945.

Korea, 1951–53.

Operation 'Killer', in which the Patricias led the Brigade advance, pursuing a retreating enemy ten miles in as many days to take a series of hills distinguished only by their spot-height numbers. As well as the enemy, the troops also had to battle against bad weather and inhospitable terrain. Hills, on average 400 metres high, had to be scaled regularly; while defensive positions had to be hacked from ground covered in snow and frozen several feet in depth. By the beginning of March all objectives had been taken, and the Patricias went into reserve.

The 27th Brigade's next advance was as part of Operation 'Ripper'. On 7 March the Patricias attacked in conjunction with the 3rd Royal Australian Regiment. Their objective was Hill 532 which, after an initial setback, was taken. During the attack Pte. L. Barton became the first Canadian to be decorated in the Korean War. He won his Military Medal for rallying his platoon after his officer had been wounded. Barton was himself wounded three times, but only went to the rear after being ordered to do so. On 13 March the 27th Brigade was relieved and moved into reserve. The

three weeks of sporadic fighting had cost the Patricias 57 casualties, 14 of them fatal.

At the end of March 1951 the 27th Brigade moved to the Kapyong Valley, where the 2nd PPCLI were to play a key part in the battle of Kapyong when the Chinese launched their spring offensive on 23–25 April. Despite the withdrawal of a neighbouring unit, which dangerously exposed a flank, and the loss of an overrun forward platoon, the Patricias clung tenaciously to their position, stopping the Chinese assault and forcing its retirement. For their gallantry the 2nd PPCLI were rewarded with a US Presidential Citation.

Meanwhile, the remaining units of the 25th Canadian Infantry Brigade Group had been completing their training at Fort Lewis, and on 21 February the decision was taken to send them to Korea. Early in May the Brigade—including the Shermans of 'C' Squadron, Lord Strathcona's Horse, the 2nd Regiment of the Royal Canadian Horse Artillery, and the second battalions of the Royal Canadian Regiment and Le Royal 22ᵉ Regiment—landed at Pusan.

The 2nd PPCLI had expected to join their comrades immediately, but the arrival of the new Canadian units coincided with a UN offensive and the 25th Brigade was diverted to support the US drive. During this operation the 2nd RCR fought a fierce action at Chail-li on 30 May, which culminated in a tricky withdrawal and cost six dead and 23 wounded. With this operation over, the Patricias were able to join their parent formation on 10 June.

The Brigade's next action was in a position between the Chorwan and Chatan valleys, from which the Canadians were employed in routine patrolling. The newly arrived men suffered as much discomfort from the climate as had the 2nd PPCLI on their arrival in Korea, but with the special miseries of the Korean summer now replacing those of winter!

Early in July truce talks began, although few realised at the time that they would drag on for two years. Also in this month, moves began for the concentration of the 1st (Commonwealth) Division; and the 25th Canadian Brigade now came under command of this formation.

For the next few months the Canadians fought a series of minor actions on the now stabilising front.

Men of the 2nd PPCLI check weapons and equipment before going into action; Korea, February 1951. (IWM KOR637)

In July and August they were occupied in extensive patrolling north of the Imjin River, and throughout September and early October in operations to reduce an enemy position threatening the UN supply route.

In the late autumn of 1951 a rotation scheme began which led to the eventual replacement of the original infantry element of the Brigade by their own 1st Battalions. The arriving units found a situation that differed little from the trench warfare of the Great War, with trenches, bunkers, wire, mines, booby-traps, artillery concentrations and machine gun fire dominating the tactical scene, and savage battles flaring as local advantages were sought. The Canadian sector was particularly active in November when the enemy attacked the recently arrived 1st PPCLI and then the veteran 2nd Royal 22ᵉ Regiment. On both occasions the Canadians stood firm against superior numbers, exacting a savage toll from the enemy.

After the November battles the tempo of the war abated to conform with negotiating postures at the truce talks, with the UN Command content to hold their existing positions and to maintain them with patrolling and defensive fire. This lull was to last right up to the armistice, broken only whenever the UN or Chinese Commands ordered the stepping-up of pressure for bargaining purposes at the negotiating table. Even so the war, and the risks run, were real enough to the soldiers on both sides, and the Canadians bore their share of the fighting and casualties in the last year-and-a-half of the conflict.

Rotation continued, and over the winter of 1952/3 the infantry saw their third relief with the arrival of the 3rd Battalions. The original Special Force element had long since disappeared, and Canada's soldiers in Korea were by now all regulars.

On 27 July 1953 armistice terms were finally agreed, and the Korean War came to an end. Some 22,000 Canadians had served in Korea or Japan during hostilities, and over 300 had died in battle; 1,200 suffered wounds, and 234 were decorated for their conduct. Canada had nobly answered the call of the UN in their first great crisis.

Units with the 25th Canadian Infantry Brigade Group, 1950–53

Lord Strathcona's Horse (Royal Canadians) (2nd Armd. Regt.): 'C', 'B' & 'A' Squadrons

Canadian infantryman, Korea.

The Royal Canadian Dragoons (1st Armd. Regt.): 'D' Squadron

2nd, 1st and 4th (81st Field) Regts., Royal Canadian Horse Arty.

2nd, 1st and 3rd Battalions of the:
    Royal Canadian Regt.
    Princess Patricia's Canadian Light Infantry
    Royal 22ᵉ Regt.
    Minor units of the Royal Canadian Engineers; Royal Canadian Signals; Royal Canadian Army Service Corps; Royal Canadian Army Medical Corps; Royal Canadian Ordnance Corps; Royal Canadian Electrical and Mechanical Engineers, and Canadian Provost Corps.

*   *   *

Post-1953: Keeping the Peace

Since the Korean armistice Canada's soldiers have been heavily engaged in the wars that have flared up around the world. Their government no longer sends them to champion this or that cause, but to maintain the fragile truces worked out under the auspices of the United Nations.

Canadian involvement with UN peacekeeping began in the wake of the Korean War when Canadian troops were sent to Kashmir, Indo-China and the Gaza strip in the 1950s. Since then they have worn the light blue beret of the UN peacekeeping forces in many parts of the world, carrying out the always difficult task of keeping the peace with the good humour and professionalism so long associated with the Canadian soldier.

And so, in the closing years of the 20th century, the Canadian Army continues to be at war—with war itself.

The Plates

A1: Late 19th-century Canadian Militiaman
The last war fought on Canadian soil was the Riel Rebellion of 1885. The Canadian Militia units that marched off to quell the rebellion did so in full dress—the only dress they had. Most units wore red coats, but the figure depicted here wears the black of the '90th Winnipeg Battalion of Rifles'. Details are taken from photographs of Sgt. T. Wright of 'F' Company, made soon after his return from active service. His weapon is a 'short' Snider rifle, .577 in.

A2: Private, 2nd (Special Service) Battalion, Royal Canadian Regiment; South Africa, 1900
Based on a well-known photograph of Canadian troops in South Africa, our figure looks very much like a Tommy of the time. The belt and frog of his Canadian Oliver equipment mark his nationality, however, as does the RCR insignia on his helmet. Other equipment includes the American-pattern Mills bandolier and a British water-bottle. His service dress is of khaki wool serge, and he is armed with the standard British .303 Lee-Enfield and bayonet of the time. This was the dress worn at the time of Paardeberg.

A3: Sergeant, Strathcona's Horse; South Africa, 1900
Raised and equipped in just over two months, Strathcona's Horse sailed for South Africa on 13 March 1900. (The title changed to Lord Strathcona's only in 1911.) There they rapidly established a reputation as an élite mounted corps. Our sergeant wears British wool serge uniform (the

Canadian issue uniforms were not suitable for South Africa), but his stetson, boots and equipment are essentially Canadian. His sidearm is a Colt .44 in. single-action Army revolver, issued to all ranks of the regiment.

B1: Sergeant, Princess Patricia's Canadian Light Infantry, 1914

Shown on arrival in England, this NCO wears the Canadian issue service dress with the insignia of his regiment. Webbing equipment is of a pattern available commercially from the Mills Equipment Company at the time; generally similar to the British official pattern, it differed in detail. On his right hip is the short-lived entrenching tool/shield

designed by Sam Hughes's secretary (an idea that proved to be a costly fiasco). His rifle is the Canadian Ross.

B2: Machine gunner, 1st Canadian Contingent, 1915

Muffled against the cold and damp of the English winter, a Canadian private carries his Colt machine gun away from the ranges. The Colt, purchased in the USA, was extensively used by the Canadians until sufficient Vickers machine guns became available to replace them. Mounted on a British tripod, it fired the standard .303 in. Canadian ammunition. Note the Canadian-pattern greatcoat.

B3: Cavalryman, 1st Canadian Contingent, 1915

Until stocks of British equipment became available to replace them, Canadian-pattern saddlery of the 'western' style was used by the cavalry of the first contingent. In the background, a trooper sits his

In December 1963 Cyprus erupted in a bloody civil war. A fragile truce was negotiated by the British troops on the spot, and the task of maintaining it was handed over to the UN contingent who began to arrive early in 1964. Here Canadian troops man a checkpoint with a sign in English, Greek and Turkish. (Canadian Defence Liaison Staff, London)

'stock' saddle protected against the English winter by a 'slicker' raincoat and stetson hat. Note the wooden stirrups and typical pommel which identified the American stock saddle.

C1: Sergeant, 73rd Battalion, CEF, 1915

Shortages of regulation clothing in the early months of the war led to the adoption of a 'khaki' tartan by certain units of the Royal Highlanders of Canada. Seen here is an NCO of the 73rd Battalion with both kilt and Glengarry cap made up in this material. Note the regimental sporran—a typical Canadian item—the Ross rifle, and the 'Oliver' pattern equipment, worn until stocks of British equipment allowed its replacement.

C2: Regimental Sergeant Major, 15th Battalion, CEF, 1918

One of the Highland regiments of the original contingent, formed from the 48th Highlanders of Canada, was the 15th Battalion. The Davidson tartan of the 48th Highlanders is visible beneath the kilt apron. Note the 'battle patch' of the battalion, indicating unit, brigade and division. Note also the blue overseas service chevrons (each indicating a year's active service), the gold wound stripes on the left cuff and the ribbons of the Military Cross and the Distinguished Conduct Medal.

C3: Lieutenant-Colonel, 16th Battalion, CEF, 1918

Lt.Col. Cyrus Wesley Peck, VC, DSO, commanded the Canadian Scottish at the close of the Great War. The details of his uniform are based on photographs taken in 1918. His breeches are cut from Regimental Mackenzie tartan, as is the 'patch' behind his bonnet badge. Note the 'battle patch'

Early days of the UN Force in Cyprus: men of the 1st Canadian Guards man an observation post in 1964. Note the .50cal. Browning M2 machine gun, and the UN patches worn on brassards and caps. (CDLS London)

worn on both arms, overseas service chevrons, and wound stripes. Three other members of the 16th Battalion won the Victoria Cross.

D1: Trooper, 19th Alberta Dragoons; England, 1916
A very 'Canadian' uniform worn during the Great War was that of the Alberta Dragoons, who retained the stetson hat throughout the war. The regimental badge is worn on the band of the hat; the ubiquitous 'CANADA' titles are worn on yellow shoulder straps (these indicated the man's sub-unit); and the Canadian-pattern Service Dress tunic is distinguishable by the seven small buttons down the front and the mitre cuffs.

D2: Driver, Canadian Field Artillery, 1918
In the closing months of the Great War men of the Canadian Corps began to paint their 'battle patches' on their helmets as well as wearing them on their sleeves; this CFA driver wears the patch of the 4th Division in this manner. Note the leather jerkin, box respirator, 'leg iron' and whip. The last two items mark our man as one who drove the teams of horses that towed the guns; the 'leg iron' prevented the driver's leg from being crushed between the off and near horses when he rode.

D3: Canadian infantryman, 1918
Sgt. William Merrifield won the Victoria Cross at Abancourt, France, on 1 October 1918. He had been in France since the first months of the war, and had been wounded and awarded the Military Medal prior to the deed that earned him the supreme award. Sgt. Merrifield's 'battle patches' are those of the 4th Battalion, CEF (Central Ontario), of the 1st Canadian Division. Details of his British-pattern uniform are taken from con-

Canadian UN Contingent in Cyprus, 1974: a mobile patrol goes about its duties in an armoured personnel carrier, clearly marked with the UN colours. (CDLS London)

temporary photographs. His rifle is the British SMLE which replaced the Canadian Ross.

E1: Lt.Col. R. O. Alexander, DSO, Royal Canadian Regiment, 1926
As the British and Empire Armies moved into an era of drab uniform, some of the splendour of the past was kept alive by the full dress retained by officers for court and levee purposes. The full dress of Canada's premier regular regiment in the 1920s is shown in this portrait taken from a contemporary photograph.

E2: Canadian infantryman; Hong Kong, 1941/42
Shown in the full battle order in which the ill-fated Canadian contingent fought the Japanese, our figure looks very similar to any British soldier in the East. Experience led to the replacement of the impractical short khaki drill trousers and shirt by a jungle-green battledress. In 1941 respirators and anti-gas capes were still an essential part of battle equipment: gas was still considered a threat to be guarded against.

E3: Canadian infantryman; Aleutian Islands, 1943
The Canadian Contingent of the joint US/Canadian task force to liberate the Aleutian Islands from the Japanese wore a mixture of Canadian and US clothing and equipment, as is evident here. Our subject's helmet, clothing and equipment are American, while his No. 4 rifle and respirator ('light, anti-gas') are Canadian. Around his waist is a lifejacket ready to be inflated in an emergency.

F1: Corporal, Fusiliers Mont-Royal; Dieppe, 1942
Pitched ashore when the battle was already lost, the Fusiliers Mont-Royal (with an embarkation strength of 584 all ranks) lost 105 killed and 354 captured on the bloody beaches of Dieppe; only 125 men returned to England. The NCO depicted here is a medical orderly, as his stretcher-bearer brassard indicates. At the time of Dieppe the 2nd Division was still using the Great War 'battle patch' system. The blue rectangle indicated the 2nd Canadian Infantry Division, while the blue circle indicated the senior battalion of the 6th Brigade.

F2: Major Paul Triquet, VC, Royal 22ᵉ Regiment; Italy, 1944
As a captain in the Royal 22ᵉ Regiment (the 'Vandoos'), Paul Triquet won the Victoria Cross on 14 December 1943 at Casa Berardi, Italy. Captain Triquet's company was reduced to less than 20 men by the time the hamlet was taken, but they held on under his leadership against repeated German counter-attacks until relieved next day. Triquet is shown as depicted in photographs taken in August 1944. The khaki drill uniform is of Canadian pattern, as are the brassards bearing regimental, national and divisional insignia.

F3: Sergeant, CWAC, 1945/46
The Canadian Women's Army Corps was formed in the autumn of 1941 and incorporated into the Canadian Armed Forces on 13 March 1942. By the end of that year nearly 10,000 CWACS were serving. More than 1,000 eventually served overseas in the Second World War. The sergeant shown wears the insignia of Canadian Military Headquarters, London. Her uniform features the 'beech-brown' and gold colours of the Corps, and her medal ribbons are those of the Canada Volunteer Service Medal and the British War Medal.

A Canadian gunner captain mans an observation post in the Middle East, 1975. He wears the tropical version of the all-services dark green uniform. Note the UN and Royal Canadian Artillery badges on his pale blue beret, rank and national insignia, and pale blue scarf. (CDLS London)

G1: Sergeant, Royal Winnipeg Rifles; 'D-Day', 1944
Dressed for the Normandy assault, our subject wears standard Canadian battledress, Mk. III helmet (by 1944 available in large quantities), a battle jerkin in lieu of conventional webbing equipment, and a light respirator. A deflated lifebelt is worn beneath the jerkin, and a shovel is carried in addition to the Sten machine carbine with which he is armed. On the sleeves of his battledress are displayed regimental titles, the patch of the 3rd Infantry Division, and regimental badges of rank.

G2: Sergeant, 12th Manitoba Dragoons, July 1944
Based on photographs of the author's uncle, 'Gerry' Bastable, who met his death in the fighting around Falaise in August 1944. The Manitoba Dragoons were the Recce Regiment for II Canadian Corps and were equipped with armoured cars. As part of the Royal Canadian Armoured Corps they wore the black beret. Note the regimental title, II Corps patch, and ribbon of the Canada Volunteer Service Medal.

G3: Brigadier, 9th Infantry Brigade, 3rd Canadian Infantry Division, 1945
The 9th Brigade consisted of three Canadian Scots Regiments: the Highland Light Infantry of Canada, the Stormont Dundas and Glengarry Highlanders, and the North Nova Scotia Highlanders. Brig. 'Rocky' Rockingham commanded this élite within an élite, always wearing Scottish headdress instead of the red-banded cap of his rank. Above the 3rd Division patch on his sleeve is the bar indicating brigade headquarters.

H1: Sergeant, Royal Canadian Regiment; Korea, 1953
Depicted on 'R and R' in Japan is an NCO of the 3rd Battalion, the Royal Canadian Regiment. His battledress is of the final (post Second World War) Canadian pattern designed to be worn with a collar and tie. A regular soldier, he wears Canadian parachutist's wings and ribbons of the Second

The setting sun throws long shadows as the white-painted jeeps of the Canadian UN Emergency Force skirt the dunes in Gaza, 1964. (CDLS London)

World War in addition to those of the Korean War. Formation signs worn are those of the 1st Commonwealth Division and the 25th Canadian Brigade.

H2: Guardsman, Canadian Guards; Cyprus 1964
One of the first Canadian contingents to the UN peacekeeping force in Cyprus, the 1st Canadian Guards discarded regimental headdress and insignia for the cap of the United Nations. Regimental shoulder titles were worn on the brassards which displayed the UN crest.

H3: Canadian Forces uniform, 1971
In the 1960s it was decided to have one uniform only for Canada's land, sea and air forces. This 'Canadian Forces' uniform began to appear in 1967 and was formalised in 1971. Our figure is a corporal of the PPCLI in ceremonial order. Note Presidential Unit Citations and the medal for UN service in Cyprus.

Notes sur les planches en couleur

A1 Homme de milice, appartenant aux *90th Winnipeg Rifles*, à l'époque de la Rébellion de Riel, 1885—la dernière 'guerre' sur le sol canadien. Les miliciens n'avaient qu'une tenue, généralement rouge mais, dans ce cas, noire. Il porte un fusil *Snider 0,577 inch*. **A2** À part les insignes de régiment sur le casque et les petits détails d'équipement, ce soldat est identique au '*Tommy*' de la même époque. **A3** Uniforme britannique, mais chapeau, bottes et équipements canadiens et revolver *Colt 0,44* porté par tous les rangs.

B1 Tenue et équipement réglementaires canadiens, l'équipement provenant de la société *Mills*, mais étant légèrement différent de celui des troupes britanniques. Sur la hanche droite, un outil de tranchée/bouclier pare-balles combiné, dispositif étrange et de brève durée inventé par la 'secrétaire' du Ministre des Milices et commandé en grandes quantités à une société dans laquelle elle avait un intérêt! L'arme est le fusil *Ross* peu fiable quoique précis. **B2** Paletot de modèle canadien, porté par un mitrailleur de *Colt*; le *Colt* était acheté aux Etats-Unis et il fut utilisé jusqu'à ce qu'un nombre suffisant de *Vickers* fut disponible pour le remplacer. **B3** Les contingents de cavalerie d'une époque précoce ont utilisé des selles et des étriers de 'cow-boys' jusqu'à ce que des articles militaires britanniques soient disponibles.

C1 Certaines unités *Highland* canadiennes portaient ce 'tartan kaki' dans les premiers mois de la guerre. Ce sergent porte aussi le fusil *Ross* et un équipement *Oliver* canadien. **C2** Formé à partir des *48th Highlanders of Canada*, le bataillon 15th Bn. porta le tartan Davidson du régiment de milice sous le revêtement de kilt traditionnel. Les insignes de manche indiquent le bataillon, la brigade (3ᵉ) et la division (1ᵉʳᵉ). Les galons bleus indiquent chacun un an de service actif, le galon or sur la manchette gauche est un 'galon de blessure'; les rubans sont ceux de la *Military Cross* et de la *Distinguished Conduct Medal*. **C3** Le *Lieutenant-Colonel C. W. Peck* était l'un de quatre membres de ce bataillon qui reçurent la *Victoria Cross*. Formé à partir du *Canadian Scottish*, le 16ᵉ bataillon portait le tartan *Mackenzie* de cette unité.

D1 Uniforme très visiblement 'canadien'; l'emblème national est porté sur les bandoulières dans la couleur de l'escadron. La tunique canadienne avait ces manchettes distinctives. **D2** Les insignes portées sur les épaules—'*battle patches*'—étaient aussi peintes sur les casques durant les derniers mois de la guerre; le '*patch*' de la 4ème Division Canadienne est représenté ici. L'attelle en fer de jambe empêche l'écrasement de la jambe de ce conducteur d'artillerie d'être écrasée entre les chevaux. **D3** D'après des photographies de Sergent W. Merrifield, VC, MM, du 4ᵉ bataillon ('Central Ontario Regt'), 1ᵉʳᵉ Brigade, 1ᵉʳᵉ Division Canadienne; notez l'uniforme britannique et le fusil Lee-Enfield.

E1 Seul l'uniforme de cérémonie conservé pur les galas s'écartait de la monotonie du kaki durant cette période. D'après une photo du commandant du 'senior regular régiment' du Canada. **E2** En tenue de bataille complète pour la résistance courte et condamnée à l'insuccès à l'invasion de Hong-Kong par les japonais, cet homme porte un uniforme et un équipement de fourniture britannique pour l'Extrême-Orient. **E3** Seuls le fusil et le masque à gaz sont canadiens; le reste de la tenue fut fourni par l'armée américaine pour ce détachement conjoint.

F1 L'écusson rectangulaire bleu indique la 2e Division, le cercle bleu indique le bataillon le plus ancien de la 6e Brigade. **F2** L'uniforme sable-kaki porté par cet homme décoré de la *Victoria Cross* est de modèle canadien, de même que le brassard qui présente les insignes du régiment, de la nation et de la division. **F3** Ce sergent féminin porte un uniforme présentant les couleurs 'marron hêtre' et or du *Canadian Women's Army Corps* et l'insigne du Quartier Général Militaire Canadien de Londres.

G1 *Battledress* canadienne (plus verte et mieux faite que la tenue britannique); casque Mk III; gilet de combat remplaçant l'équipement personnel ordinaire; gilet de sauvetage et respirateur. Les insignes comprennent le '*title*' du régiment à la partie supérieure de la manche, l'écusson de la 3e Division et les insignes de rang du type du régiment. **G2** Béret noir du *Royal Canadian Armoured Corps*; '*title*' du régiment et insignes du *II Corps*. **G3** Le brigadier 'Rocky' Rockingham porte un bonnet écossais (la 9e brigade était entièrement composée d'unités canadiennes-écossaises), l'écusson de la 3e division et le galon indiquant le quartier général de la brigade.

H1 En permission au Japon, ce sous-officier porte le modèle final de la *battledress* canadienne avec les insignes de la 25e brigade canadienne et de la 1ᵉʳᵉ division du Commonwealth, ses ailes de parachutiste et certains de ses rubans de médaille datent de la deuxième guerre mondiale. **H2** La coiffe est celle de l'ONU ainsi que l'insigne de bras sous le '*title*' d'épaule du régiment. **H3** L'uniforme des années 1970, porté par les trois forces armées du Canada—ici par un caporal de la *Princess Patricia's Canadian Light Infantry* en tenue de parade. Notez l'insigne de *US Presidential Unit Citation*, obtenue en Corée.

Farbtafeln

A1 Milizsoldat der *90th Winnipeg Rifles* zur Zeit des Riel-Aufstands im Jahre 1885, des letzten 'Kriegs' auf kanadischem Boden. Milizeinheiten hatten lediglich ihre grosse Dienstuniform: weitgehend rot, aber in diesem Fall schwarz. Dieser Vertreter trägt ein *Snider* Gewehr mit *0,577 inch* Kaliber. **A2** Abgesehen von dem Regimentsabzeichen auf dem Helm und Einzelheiten in der Ausrüstung ist dieser Soldat identisch mit dem '*Tommy*' der gleichen Epoche. **A3** Britisch Uniform und kanadischer Hut sowie Stiefel und Ausrüstung; alle Ränge hatten Colt .44 Revolver.

B1 Kanadische Standarduniform und-ausrüstung, letztere von der Firma *Mills* erworben, aber etwas anders als die britische Ausführung. An der rechten Hüfte eine Verbindung aus Schützengrabenschanze und Kugelschutzschild. Die auffällige, kurzlebige Einrichtung, die von der '*Sekretärin*' des Milizministers erfunden und in grossen Mengen bei einer Firma bestellt wurde, bei der diese Dame Teilhaberin war; die Waffe ist ein unzuverlässiges, aber exaktes *Ross* Gewehr. **B2** Gemusterter kanadischer Mantel, von den *Colt* MG-Schützen getragen; das *Colt* MG wurde in den USA gekauft und verwendet, bis ausreichend *Vickers* Gewehre vorlagen, um es zu ersetzen. **B3** Frühe Kavallerieeinheiten benutzten 'Cowboy'-Sättel und Zaumzeug, bis die britischen Militärausführungen erhältlich waren.

C1 Einige kanadische *Highland*-Einheiten trugen dieses khakifarbene Tartanmuster in den ersten Monaten des Krieges. Dieser Sergant hat ausserdem ein Ross Gewehr und kanadische *Oliver* Ausrüstung. **C2** Das 15. Bataillon wurde aus der Einheit der *48th Highlanders of Canada* gebildet und den *Davidson* Tartan dieses Milizregiments unter dem üblichen Kilt. Die Ärmelabzeichen verweisen auf Bataillon, Brigade (3) und Abteilung (1). Die blauen Winkel markieren je ein aktives Dienstjahr, der goldene Streifen auf der linken Manschette bezeichnet eine Verwundung; die Bänder gehören zum *Military Cross* und der *Distinguished Conduct Medal*. **C3** Lt.Col. *C. W. Peck* war einer von vier Mitgliedern dieses Bataillons, die das *Victoria Cross* errangen. Das aus den *Canadian Scottish* gebildete 16. Bataillon trug den *Mackenzie* Tartan dieser Einheit.

D1 Eine sehr spezifisch kanadische Uniform; der Name des andes erschein in Staffelfarben auf den Schulterstreifen. Die auffälligen Manschetten sind typisch für die kanadische Uniformjacke. **D2** Die auf den Schultern getragenen Abzeichen ('*battle patches*') wurden in den letzten Monaten des Krieges auch auf den Helmen aufgemalt; dies ist der '*patch*' der 4th Canadian Division. Der eiserne Beinschutz bewahrt die Beine dieses Artilleriefahrers davor, zwischen den Pferden dieser Gruppe zermalmt zu werden. **D3** Eine Zeichnung nach Fotos von Sgt. W. Merrifield, VV, MM vom 4. Bataillon (*Central Ontario Regt.*) 1st Bde, 1st Canadian Division; man beachte die britische Uniform und das Lee-Enfield Gewehr.

E1 Nur die bei Gala-Anlässen getragene volle Ausgehuniform brachte eine Abwechslung von den düsteren Khakifarben dieser Periode. Nach einem Foto des Oberbefehlshabers des ältesten kanadischen regulären Regiments. **E2** Dieser Mann trägt die vollständige Schlachtausrüstung für den kurzen und zwecklosen Widerstand gegen die japanische Invasion in Hong Kong, mit britischer Fernostuniform und -ausrüstung. **E3** Nur Gewehr und Gasmaske stammen aus Kanada, die restliche Ausrüstung wurde von der US Army für diese gemeinsamen Streitkräfte zur Verfügung gestellt.

F1 Das blaue, rechteckige Abzeichen verweist auf die 2nd Division, der blaue Kreis auf das ältestes Bataillon der 6th Brigade. **F2** Die von diesem Träger des *Victoria Cross* getragene sand- und khakifarbene Uniform ist von kanadischem Schnitt, ebenso wie die Armbinde mit Regiments-, National- und Abteilungsabzeichen. **F3** Dieser weibliche Sergant trägt eine Uniform in den birkenbraunen und goldenen Farben des Canadian Women's Army Corps und die Abzeichen des kanadischen Militärhauptquartiers in London.

G1 Kanadisches '*battledress*' (grüner und besser geschniedert als die britische Ausführung), *Mk III* Helm, 'Schlachtjacke' anstelle der konventionellen individuellen Ausrüstung, Schwimmweste und Atemgerät. Zu den Abzeichen gehört der Regiments-'*title*' oben auf dem Ärmel, das Zeichen der 3rd Division und Rangabzeichen mit dem Regimentsmuster. **G2** Schwarzes Barrett des *Royal Canadian Armoured Corps*, Regiments-'*title*' und *II Corps* Abzeichen. **G3** Brigadier 'Rocky' Rockingham trägt die schottische Kappe (die 9th Brigade bestand ausschliesslich aus kanadisch-schottischen Einheiten), das Abzeichen der 3rd Division und den Streifen für das Brigadehauptquartier.

H1 Dieser Unteroffizier auf Urlaub in Japan trägt das letzte Muster des kanadischen *battledress* mit den Abzeichen der 25th Canadian Brigade und der 1st Commonwealth Division; die Fallschirmspringer-'Flügel' und einige seiner Metallbänder stammen aus dem Zweiten Weltkrieg. **H2** Die Kopfbedeckung stammt von der UNO, ebenso wie die Armabzeichen unter dem Regiments-'*title*' auf der Schulter. **H3** Die Uniform der 1970er Jahre, von allen drei kanadischen Waffengattungen getragen, in diesem fall handelt es sich um einen Korporal der *Princess Patricia's Canadian Light Infantry* in Paradeuniform. Man beachte die in Korea erworbene *US Presidential Unit Citation* Abzeichen.